VITAL
SIGNS
2005

Other Norton/Worldwatch Books

VITAL SIGNS

2005

**The Trends That Are
Shaping Our Future**

WORLDWATCH INSTITUTE

Lisa Mastny, *Project Director*

Molly Aeck

Erik Assadourian

Zoë Chafe

Christopher Flavin

Hilary French

Gary Gardner

Brian Halweil

Nicholas Lenssen

Danielle Nierenberg

Michael Renner

Janet Sawin

Howard Youth

Linda Starke, *Editor*

Lyle Rosbotham, *Designer*

W.W. Norton & Company
New York   London

Worldwatch Institute Staff

Contents

PART TWO: **Special Features**

Acknowledgments

"Some books are to be tasted, others to be swallowed, and some few to be chewed and digested," wrote English essayist Francis Bacon. We hope that you, our readers, will take the time to thoroughly chew and digest the information contained in *Vital Signs 2005*. It is through the support of readers like you that we have been able to make this book available to educators, activists, journalists, government officials, and others around the world.

In British Columbia, Canada, for instance, a professor of sociology uses the 50-year time series data in *Vital Signs* to establish trends in his own research and writing, while a local Green Party official uses the analysis to inform his political commentary in the media. In Spain, the Advisory Council for the Sustainable Development of Catalonia distributes *Vital Signs* to all council members, local government officials, and presidents of Catalan universities to promote awareness of environmental protection and quality-of-life issues. And in Alaska, the owners of a small eco-lodge make *Vital Signs* available to their guests because it "provides… objective information free of consumption-driven marketing 'spin'." We also appreciate the support of print, radio, and television journalists across the globe who rely on the volume as a research and reference tool.

In checking the planet's vital signs, we depend on numerous experts who kindly donate their time to comment on drafts or provide the data that we rely on to write each piece. For all the help we received this year, we especially thank Howard Cambridge, Richard Cincotta, Colin Couchman, Brigitte Du Jeu, Torbjörn Fredriksson, Lew Fulton, Paul Gipe, Claudia Grotz, Wilfried Haeberli, Lotta Harbom, Steven Hedlund, Martha Honey, Alan Lopez, Birger Madsen, Paul Maycock, Corin Millais, Sara Montanaro, Miquel Muñoz, James Paul, Christine Real de Azua, David Roodman, Marc Sani, Wolfgang Schreiber, Vladimir Slivyak, Werner Weiss, Jessica Wenban-Smith, John Whitelegg, Tim Whorf, and Angelika Wirtz.

At our longtime publisher, W.W. Norton & Company, we are fortunate to work with Amy Cherry, Lucinda Bartley, and Leo Wiegman. With their help, *Vital Signs* is transformed from a jumble of documents and data files into this volume that is found in bookstores and classrooms across the United States.

We are also lucky enough to have a committed group of international partners who are interested in getting *Vital Signs* published outside the United States in many languages. For their considerable help in publishing and promoting recent editions, we thank Soki Oda of Worldwatch Japan, Anna Bruno Ventre of Edizioni Ambiente in Italy, Gianfranco Bologna of WWF Italy, Sang Baek Lee and Jung Yu Jin of the Korean Federation for Environmental Movement, Lluis Garcia Petit and Sergi Rovira at Centro UNESCO de Catalunya in Spain, Marisa Mercado at Fundación Hogar del Empleado in Spain, Eduardo Athayde of UMA–Universidade Livre da Mata Atlantica in Brazil, Eilon Schwartz of the Heschel Center for Environmental Learning and Leadership in Israel, and Hamid Taravati in Iran.

Worldwatch's general research program is supported each year by numerous philanthropic organizations that are concerned about the state of the world. Without their support, we would be unable to track these vital signs. We thank the following foundations for their generous support over the last year: Aria Foundation; the Blue Moon Fund; GTZ, the German Society of Technical Co-operation; Goldman Environmental Prize/Richard and Rhoda Goldman Fund; The William and Flora Hewlett Foundation; the W.K. Kellogg Foundation; the Frances Lear Foundation; the Steven C. Leuthold Family Foundation; the Massachusetts Technology Collaborative; the Merck Family Fund; the Norwegian Royal Ministry of Foreign Affairs; The Overbrook Foundation; the V. Kann Rasmussen Foundation; the Rockefeller Brothers Fund; The Shared Earth Foundation; The Shenandoah Foundation; the Turner Foundation, Inc.; the U.N. Population Fund; the Wallace Genetic Foundation, Inc.; the Wallace Global Fund; the Johanette Wallerstein Institute; and The Winslow Foundation.

The Institute is also supported by thousands of Friends of Worldwatch, whose commitment to the future of the Institute has made our work possible. Special thanks go to our Council of Sponsors: Adam and Rachel Albright, Tom and Cathy Crain, John and Laurie McBride, and Kate McBride Puckett. And we thank Worldwatch's Board of Directors, an exceptional group of people whose commitment and leadership over the last year continue to guide the Institute in rapidly changing times.

Beyond the authors whose names you will find on individual vital signs, Worldwatch relies on a professional staff who are equally committed to making progress toward a sustainable society. Patricia Shyne, in charge of business development, works closely with W.W. Norton and our international partners. Our development team of John Holman, Mary Redfern, and Mairead Hartmann is expanding Worldwatch's broad base of support, while our communications team of Darcey Rakestraw and Courtney Berner brings *Vital Signs* and other Worldwatch publications to new audiences every day. Backing us all up are Barbara Fallin, our Director of Finance and Administration; Web Manager Steve Conklin; and Joseph Gravely in our mailroom.

We are fortunate to be able to draw on not only current researchers but former Worldwatchers as well. Alumni Howard Youth and Nick Lenssen contributed three pieces this year. Research Fellow Eric Martinot provided valuable insights on renewable energy trends and technologies. Helping researchers old and new is the task ably performed by Research Librarian Lori Brown. And in addition to writing their own vital signs, Staff Researchers Erik Assadourian and Zoë Chafe and Renewable Energy Program Manager Molly Aeck were especially helpful in providing background research to Hilary French and

Janet Sawin on several pieces.

Last but not least, for each edition of *Vital Signs* our editor and our art director turn discrete pieces of prose, data tables, and graphs into this coherent book—even under escalating deadline pressure. Independent editor Linda Starke cracks the whip to get authors to deliver their texts on time. And this year Art Director Lyle Rosbotham took the bold step of adding color to many design elements. He also finds all the photos that anchor the vital signs in the real world and bring a human face to many of the trends we document.

We hope that *Vital Signs* this year will give you something to chew on, as Francis Bacon put it. And perhaps it will give you ideas for your own vital signs. To help you develop those, the data used to prepare all the graphs in this book are available on our CD-ROM, *Signposts*. Do let us know if you have ideas of trends we should cover in future editions. You can reach us by e-mail (worldwatch@worldwatch.org), fax (202-296-7365), or regular mail.

Lisa Mastny
March 2005

Worldwatch Institute
1776 Massachusetts Ave., N.W.
Washington, DC 20036

Preface

The year 2004 was a record breaker virtually across the board. The world economy expanded at a scorching 5 percent rate, pushing consumption and production of everything from grain and meat to steel and oil to new highs. These burgeoning physical indicators of growth are powerful reminders that despite the popular image, our "post-industrial" information age has by no means freed itself from the material world.

Steel, the archetypal twentieth-century industrial metal, is a case in point. World steel production jumped by a remarkable 9 percent in 2004, crossing the billion-ton threshold for the first time. This is 33 percent above the level of world production just five years earlier, and marks a dramatic acceleration from the growth rates in the 1980s and 1990s. Traditionally an industry centered in the northern industrial powers, it is also striking that the United States accounted for less than 10 percent of steel output in 2004.

The explanation for these sudden shifts is summed up in a single word: China. In the case of steel, China's production has more than doubled in the last four years and now accounts for 27 percent of the world total—140 percent above the second-place producer, Japan. China's remarkable levels of steel production and consumption reflect the fact that this country is entering a new stage of economic development, one that requires a massive expansion in its limited physical infrastructure, from roads to factories and buildings.

With its relatively modest endowment of natural resources, China is now using its massive foreign exchange earnings from manufacturing to draw in resources from around the globe. In terms of scale, this is as if all of Europe, Russia, North and South America, and Japan were to simultaneously undertake a century of economic development in a few short decades. And many other parts of the so-called developing world are moving nearly as rapidly in the same direction—starting particularly in East Asia, but with India and other South Asian economies also beginning to shift into higher gear.

Food markets are one place where growth in China and elsewhere is changing the landscape. The global grain harvest shot up by 8 percent to over 2 billion tons in 2004, driven by rising consumption and changing diets. Production of meat and fish—the latter increasingly derived from fish farms—also hit new highs. Grain reserves remained near historically low levels at the end of the year, leaving the world vulnerable to higher prices should the 2005 harvest be hurt by adverse weather conditions.

Although the rise in food harvests in 2004 was aided by unusually favorable weather in key countries, it was also made possible by an increase in the area cultivated, a trend that cannot continue for long without running into severe ecological constraints. Water shortages in many regions will almost certainly force a reduction in cultivated area in the years ahead. And in Brazil, the accelerated expansion of agriculture into the Amazon Basin puts at risk a region of immense and fragile biodiversity.

While convulsions in the food system may lie ahead, world oil markets have already entered their most turbulent period in more than two decades. Surging demand caused oil prices to double to a peak of $55 per barrel. By the second half of 2004, many news organizations were reporting the price of oil daily, along with stock market averages. Evidence is beginning to accumulate that there is simply not enough readily available oil to sustain current rates of demand growth. The consequent collision between demand and supply could make 2004 oil prices look like a mild warm-up to a more dramatic shock to the global economy.

From Africa to South America, Chinese and Indian companies are now competing with American and European firms for access to the few remaining frontiers of the world oil industry. This struggle for supplies is likely to intensify in the next few years. The biggest losers will be countries that have virtually no impact on the world oil market—poor oil-importing nations in Africa, Asia, and Latin America.

As global oil reserves dwindle, markets for new energy technologies will likely boom—and in 2004, some already did. Dramatic growth surges drove up production of wind turbines, solar cells, solar hot water systems, and biofuels derived from crops and agricultural wastes. Averaged over the last five years, total use of solar and wind energy is expanding at a 30-percent annual rate—doubling every three years.

Although they are just beginning to establish themselves in the new energy technology markets, China and India could have a huge impact in the next few years. With limited domestic reserves of oil and gas, strong manufacturing sectors, and an abundance of skilled and low-cost workers, China, India, and other developing countries are well positioned to claim the leadership positions in new and renewable energy now held by Europe and Japan. If they adopt the policy reforms needed to be successful with these new technologies, developing countries will drive costs down and open up a potential expressway to a post-petroleum economy.

Even so, recent developments suggest that the world economy will be coping with the downsides of its voracious material appetite for decades to come. Total carbon emissions and atmospheric concentrations of carbon dioxide are both accelerating, and 2004 was the fourth warmest year ever recorded—in fact, the 10 warmest years in the last 120 have all occurred since 1990.

Rising temperatures are rapidly melting polar ice caps and mountain glaciers around the globe. According to the Arctic Climate Impact Assessment released in 2004 by scientists from countries with Arctic territories, the rapid rise in polar temperatures is thinning the ice at a pace that could make the Arctic Ocean ice-free by 2100. While melting ocean ice does not raise the sea level, the collapse of the ice sheets covering Greenland and Antarctica will. In the last three decades, 13,500 square kilometers of Antarctic ice shelves have already disintegrated, and two of the largest ice sheets have begun to weaken.

The world's ecological systems are already in trouble, as a host of human forces impinge on coral reefs, tropical forests, and other critical natural systems. Among the canaries in this ecological coal mine are our fellow mammals, nearly one quarter of which are in serious decline. On another front, half the world's wetlands are already gone. The importance of wetlands and another key ecosystem—coral reefs— was driven home by the late 2004 tsunamis, the effects of which would have been ameliorated, scientists say, if wetlands had not been severely damaged by development.

While the living standards of many human beings have improved in the last year as incomes have grown, the rising economy did not lift all boats. The surging economy was good news for investors as profit margins revived and stock markets moved upward. But many workers did not do as well, with unemployment rates remaining high in many countries and with personal incomes remaining stagnant except for those near the top of the economic pyramid.

Those at the bottom are not doing well at all. Some 852 million people go hungry each day, according to a 2004 estimate, equivalent to the combined populations of North America, Japan, and Europe—an increase of 18 million over the last decade. Even more people lack access to clean water and sanitation. The burden of infectious diseases ranging from malaria to cholera is also growing for many of these peo-ple. The number of HIV-infected people rose to almost 78 million in 2004, nearly double the 1997 total. And the devastating Indian Ocean tsunamis that killed more than 150,000 people at the end of 2004 underscored the vulnerability of living conditions in poorer countries.

In recognition of such problems, the world community committed in September 2000 to the Millennium Development Goals, which are intended to reduce poverty rates by focusing on goals such as providing primary education for all children, empowering women, and reducing disease rates. Progress has been made on some of these goals, but the advances are uneven due to political failures in poor countries and short-falls in promised assistance from wealthier nations. Military expenditures have increased since 2000, making it difficult to meet domestic and international commitments on water sup-ply, education, and health care.

In short, the world is in the midst of a period of unprecedented and disruptive change, offering enormous opportunities and even greater risks. We hope that *Vital Signs 2005* will help people see and understand some of the big-picture trends that are too often ignored in daily news reports. Understanding the dynamic present is a first step, we believe, to creating a better future.

Christopher Flavin
President
Worldwatch Institute

TECHNICAL NOTE

Units of measure throughout this book are metric unless common usage dictates otherwise. Historical population data used in per capita calculations are from the Center for International Research at the U.S. Bureau of the Census. Historical data series in *Vital Signs* are updated in each edition, incorporating any revisions by originating organizations.

Unless otherwise noted, references to regions or groupings of countries follow definitions of the Statistics Division of the U.N. Department of Economic and Social Affairs.

Data expressed in U.S. dollars have for the most part been deflated to 2003 terms. In some cases, the original data source provided the numbers in deflated terms or supplied an appropriate deflator, as with gross world product data. Where this did not happen, the U.S. implicit gross national product (GNP) deflator from the U.S. Department of Commerce was used to represent price trends in real terms.

VITAL
SIGNS

2005

Part One

KEY INDICATORS

Food Trends

North Sea fisherman releasing catch on deck

▶ Grain Harvest and Hunger Both Grow

▶ Meat Production and Consumption Rise

▶ Aquaculture Pushes Fish Harvest Higher

Grain Harvest and Hunger Both Grow

Brian Halweil

Farmers reaped a record grain harvest of 2,049 billion tons in 2004.[1] (See Figure 1.) This haul was 9 percent above the 2003 harvest, and it broke 2 billion tons for the first time in history. Corn, wheat, and rice account for 85 percent of the global grain harvest (in terms of weight), with sorghum, millet, barley, oats, and other less common grains rounding out the total.[2]

The bumper crop of grains, which provide nearly half of humanity's calories, pushed the harvest per person to 322 kilograms—nearly 8 percent above the previous year, but still 6 percent below the peak of 343 kilograms in 1985.[3] (See Figure 2.)

pp. 24, 38

The global corn harvest hit 705 million tons in 2004.[4] Since 2001, it has surpassed that of wheat and rice as growing demand for meat encourages farmers to plant corn as a feedgrain. In both Argentina and Brazil, for instance, corn production has nearly doubled since 1990, to 13 million tons and 42 million tons respectively.[5]

Major corn-growing areas in North America, Europe, and Asia all saw bumper crops in 2004. American corn farmers, who account for over 40 percent of global production, pulled in nearly 298 million tons, smashing the previous year's record by 16 percent.[6]

After declining harvests in five of the last seven years, wheat farmers also reaped a record—624 million tons, 11 percent above the 2003 harvest.[7] In 2004, China, the world's largest wheat producer and consumer, also became the largest wheat importer despite a 6-percent increase in the national harvest.[8] Growing harvests in Argentina, Australia, and former Soviet states have reduced the U.S. share of world wheat exports from about 33 percent a decade ago to 25 percent today.[9]

The United Nations declared 2004 the international year of rice, an unprecedented distinction for a food, to raise awareness about a "pending crisis" as rice demand outpaces production.[10] More than half of the world's people eat rice as their major staple.[11] Rice provides 20 percent of the world's dietary energy supply in terms of calories consumed directly, while wheat supplies 19 percent and corn 5 percent.[12]

Although rice farmers did manage to boost the harvest in 2004, demand exceeded production for the fourth consecutive year, and tight supplies in big exporters like China, India, and Pakistan pushed prices higher.[13] As a result, global rice stocks dipped by 17 percent from 2003, to their lowest level since 1984.[14] The world's governments currently store enough rice to last just 63 days.[15]

Still, the bumper harvest for wheat and corn helped push global grain stocks up for the first time in five years. (See Figure 3.) They currently stand at 441 million tons, equivalent to about 80 days of consumption.[16] The United States and Europe enjoyed much of this increase, since the collective harvest in wealthy nations jumped 15 percent, while developing nations pulled in 3 percent more than last year.[17] In addition, a long-term slide in Chinese grain stocks, which had helped reduce the level of stocks worldwide for the last few years, slowed in 2004.[18]

The record output worldwide resulted in part from farmers planting more land in grains.[19] In 2004, the world's grain farmers worked 681 million hectares of land, the highest level since 1997.[20] Yet this was still more than 45 million hectares (6 percent) below the historic high in 1981.[21]

A more powerful engine for the record grain output was a boost in average yields, the amount of grain harvested per hectare. For the first time, grain yields in 2004 surpassed 3 tons—nearly three times the level in 1960.[22] Near-perfect weather in major growing areas helped farmers raise the yield, as fertilizer use and irrigation remained stable.[23]

Nonetheless, for the first time since it began keeping track in the 1970s, the U.N. Food and Agriculture Organization reported that the number of hungry people around the world increased.[24] Some 852 million people go hungry each day, about 18 million more than during the mid-1990s.[25] Most people go hungry not because of a global shortage of food but because they are too poor to buy food or to get the land, water, and other resources needed to produce it.[26] Hunger now kills more than 5 million children each year—roughly one child every five seconds.[27]

Figure 1. World Grain Production, 1961–2004

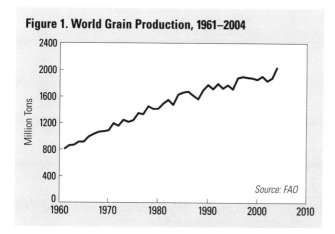

Figure 2. World Grain Production Per Person, 1961–2004

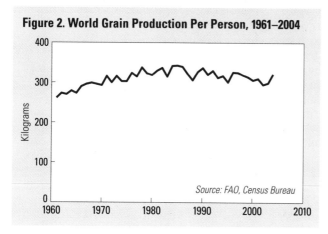

Figure 3. Grain Stocks in Industrial and Developing Countries, 1961–2005

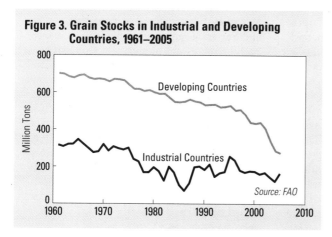

World Grain Production, 1961–2004

Year	Total	Per Person
	(million tons)	(kilograms)
1961	805	261
1962	858	273
1963	867	270
1964	914	279
1965	914	273
1966	992	290
1967	1,032	296
1968	1,065	299
1969	1,073	296
1970	1,087	293
1971	1,194	316
1972	1,156	300
1973	1,246	316
1974	1,216	303
1975	1,241	303
1976	1,348	324
1977	1,333	315
1978	1,454	338
1979	1,413	323
1980	1,418	319
1981	1,496	330
1982	1,552	337
1983	1,478	315
1984	1,632	342
1985	1,665	343
1986	1,678	340
1987	1,618	322
1988	1,565	306
1989	1,700	327
1990	1,779	337
1991	1,717	320
1992	1,797	330
1993	1,727	312
1994	1,777	317
1995	1,715	301
1996	1,883	326
1997	1,903	325
1998	1,891	319
1999	1,882	314
2000	1,860	306
2001	1,909	310
2002	1,839	295
2003	1,884	299
2004 (prel)	2,049	322

Source: FAO and U.S. Bureau of the Census.

Meat Production and Consumption Rise   Danielle Nierenberg

Worldwide meat production continues to grow, with an estimated 258 million tons produced by farmers in 2004, a 2-percent increase from 2003.[1] (See Figures 1 and 2.) Since the 1970s, meat production has more than doubled because of higher demand and the introduction of large-scale production processes.[2]

Consumption, especially in the developing world, continues to rise as well, with the average person eating almost 30 kilograms of meat a year; in industrial nations, people eat about 40 kilograms of meat annually.[3]

The International Food Policy Research Institute estimates that by 2020 people in developing countries will eat more than 36 kilograms of meat on average—twice as much as in the 1980s.[4] In China, the figure is expected to be 73 kilograms, a 55-percent increase from 1993, while in Southeast Asia people are likely to be eating 38 percent more meat than they do now.[5] People in industrial countries, however, will still consume the most—nearly 90 kilograms a year by 2020, the equivalent of a side of beef, 50 chickens, and one pig.[6]

LINKS pp. 22, 26

Global poultry output stood at 77.2 million tons in 2004, up only 1.6 percent from the previous year.[7] (See Figure 3.) This was the slowest growth ever, partly due to the widespread outbreak of avian influenza in Asia, which killed more than 40 people and forced the slaughter of some 200 million chickens.[8] As a result, production in Asia was down by 3 percent.[9] In contrast, poultry production was up 7 percent in South America.[10] Despite stiff competition from Brazil, the United States remains the world's largest producer and consumer of poultry.[11]

Beef production rose by less than 1 percent and global beef trade declined by more than 6 percent because of bans on imports from North America after the first reported cases of mad cow disease in the United States and Canada in 2003 and 2004.[12] Healthier livestock and the absence of the United States from the market pushed meat exports in South America up by 30 percent in 2004.[13]

Pork production reached more than 100 million tons in 2004 as demand grew for alternative meats in part due to public concern over avian flu and mad cow disease.[14] Developing countries, especially in Asia, accounted for more than 60 percent of global pork production in 2004, up more than 50 percent from a decade ago.[15] Yet annual consumption of pork is still low in developing nations, at 12.3 kilograms per person, compared with 30 kilograms in industrial countries.[16]

As production and consumption of meat continue to increase worldwide, the methods of production are also changing. Industrial animal agriculture, or "factory farming," is the most rapidly growing production system for pigs, chickens, and beef. More than half of the world's poultry and pork and much of the beef is produced in these intensive, inhumane, and potentially hazardous conditions.[17]

These farms also require extensive inputs—producing 8 ounces of beef requires 25,000 liters of water, for instance.[18] A calorie of beef takes 33 percent more fossil fuel to produce than a calorie of energy from potatoes would.[19] And 95 percent of the world's global soybean harvest is consumed by animals, not people.[20] In addition, cows, pigs, and chickens get 70 percent of the antimicrobial drugs produced in the United States.[21]

As environmental and public health concerns about meat production and consumption grow, farmers, business owners, chefs, and consumers are beginning to think differently about their food choices. For example, in 2003 McDonald's—the largest U.S. purchaser of beef and one of the largest buyers of chicken and pork—asked some suppliers to stop using antibiotic growth promoters in animal feed.[22] And Whole Foods Market, a Texas-based natural and organic foods supermarket, has committed $500,000 to establishing a foundation to study humane animal farming methods.[23]

Consumers are also demanding more grass-fed meat, milk, and eggs for health reasons—grass-fed products are higher in Omega 3 fatty acids, which help lower cholesterol, and in conjugated linoleic acid, which can block tumor growth and lower the risk of obesity and other diseases.[24]

Figure 1. World Meat Production, 1961–2004

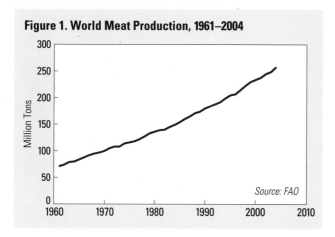

Source: FAO

Figure 2. World Meat Production Per Person, 1961–2004

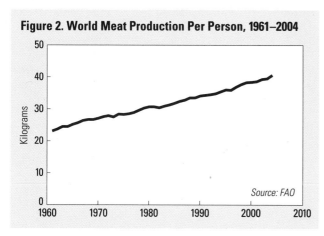

Source: FAO

Figure 3. World Meat Production by Source, 2004

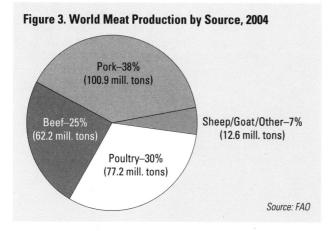

Pork–38%
(100.9 mill. tons)

Beef–25%
(62.2 mill. tons)

Sheep/Goat/Other–7%
(12.6 mill. tons)

Poultry–30%
(77.2 mill. tons)

Source: FAO

World Meat Production, 1961–2004

Year	Total	Per Person
	(million tons)	(kilograms)
1961	71	23.1
1962	74	23.7
1963	79	24.5
1964	80	24.5
1965	84	25.2
1966	88	25.7
1967	92	26.4
1968	95	26.7
1969	97	26.7
1970	100	27.1
1971	105	27.6
1972	108	27.9
1973	108	27.5
1974	114	28.4
1975	116	28.3
1976	118	28.5
1977	122	28.9
1978	127	29.6
1979	133	30.3
1980	136	30.7
1981	139	30.7
1982	140	30.4
1983	145	30.9
1984	149	31.3
1985	154	31.8
1986	160	32.4
1987	165	32.8
1988	171	33.5
1989	174	33.5
1990	180	34.1
1991	184	34.3
1992	188	34.5
1993	192	34.8
1994	199	35.4
1995	205	36.0
1996	207	35.9
1997	215	36.9
1998	223	37.7
1999	230	38.3
2000	234	38.4
2001	238	38.6
2002	245	39.3
2003	249	39.5
2004 (prel)	258	40.6

Source: FAO.

Aquaculture Pushes Fish Harvest Higher

Brian Halweil

The world's fishers harvested 133 million tons of fish and shellfish from streams, oceans, and other water bodies in 2002, the most recent year for which data are available.[1] (See Figure 1.) This record haul was nearly 2 percent more than in 2001, and nearly seven times the global harvest in 1950.[2] Over that same period, the amount of fish harvested per person tripled to 21 kilograms per year.[3] (See Figure 2.)

Asia's fishing fleet and fish farmers pulled in 81 million tons, or just over 60 percent of the world harvest.[4] China alone harvested 46 million tons, more than one third of the global total.[5] Asia is home to three of the top five fish-producing nations: China, India (6 million tons), and Indonesia (5.4 million tons).[6] Peru and the United States round out the top five, respectively harvesting 8.8 million tons and 5.4 million tons.[7]

pp. 24, 90

Marine areas yield 100.4 million tons of fish and shellfish, with anchovy, pollack (a type of cod), and tuna topping the list of species caught.[8] (See Figure 3.) Freshwater fishing accounts for the remaining 32.6 million tons, more than half of which is carp and almost all of which is harvested in developing nations.[9] (In wealthy nations, freshwater fishing is primarily a recreational activity.)[10]

Until the mid-1980s, vessels from wealthier nations dominated the ocean catch. But "exclusive economic zones," which gave all nations control over nearby waters, and the spread of industrial fishing technology helped tip the balance.[11] Today, fishers from developing countries catch three out of four wild fish (by weight).[12]

People in the developing world also eat most of the world's fish, although they consume much less per capita: 14.2 kilograms a year compared with 24 kilograms in the industrial world. For nearly 1 billion people, mostly in Asia, fish supply 30 percent of protein; worldwide, the figure is just 6 percent.[13]

Nonetheless, seafood trade tends to flow from poorer to wealthier nations, who purchased 82 percent of the $61 billion of seafood imports in 2002.[14] Shrimp alone accounts for 20 percent of global seafood trade.[15] The developing world makes more money from seafood than from coffee, cocoa, tea, or any other agricultural commodity.[16] Unfortunately, many of the 200 million people who depend on fisheries for a living—fishing families, boat builders, fishmongers—cannot afford to eat the fish they catch and handle.[17]

Fish are the last wild meal in the human diet.[18] But as more vessels work a limited number of fisheries, roughly two thirds of the world's major stocks are now fished at or beyond their capacity, and another 10 percent have been harvested so heavily that fish populations will take years to recover.[19] In 2004, marine scientists estimated that industrial fleets have fished out at least 90 percent of all large ocean predators—tuna, marlin, swordfish, sharks, cod, halibut, skates, and flounder—in just the past 50 years.[20]

With the depletion of wild fish schools, virtually all growth in the global catch today comes from farmed fish. The aquaculture harvest has doubled in the last decade, to 39.8 million tons, and now accounts for 30 percent of the global fish harvest. By 2020, it could produce nearly half of all fish harvested.[21] In China, which raises 70 percent of the world's farmed fish, this category already accounts for nearly two thirds of total fish production.[22] For some species, like salmon, farmed production now surpasses the wild harvest.[23]

In recent years, an explosion of ecolabels has helped shoppers support healthier fishing practices. Today 225 Marine Stewardship Council–labeled products are available in 22 countries. Ten fisheries have already earned certification under this program, which sets standards for responsible fishing practices, and 16 more are being assessed.[24]

An even more ambitious effort to help understand and preserve fisheries is the Census of Marine Life, a 10-year project that involves hundreds of scientists in 70 nations. In 2004, the census revealed more than two new species of fish a week and mapped vast transoceanic migration routes for turtles, tuna, and other sea life.[25]

Figure 1. World Fish Harvest, 1950–2002

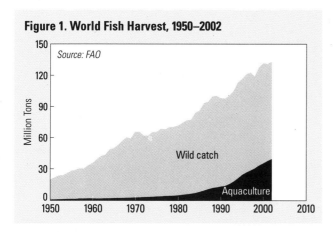

Source: FAO

Wild catch

Aquaculture

Figure 2. World Fish Harvest Per Person, 1950–2002

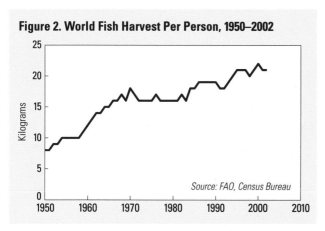

Source: FAO, Census Bureau

Figure 3. Top Fish Species Harvested Worldwide, 2002

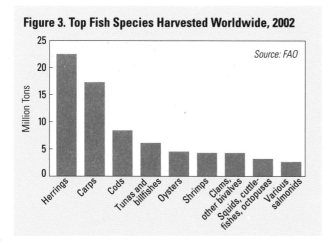

Source: FAO

World Fish Catch and Aquaculture, 1950–2002

Year	Catch	Aquaculture
	(million tons)	
1950	19	0.6
1955	27	1.2
1960	34	1.7
1965	48	2.0
1970	63	2.6
1971	63	2.7
1972	59	3.0
1973	59	3.1
1974	62	3.3
1975	62	3.6
1976	65	3.7
1977	64	4.1
1978	66	4.2
1979	66	4.3
1980	67	4.7
1981	69	5.2
1982	71	5.7
1983	71	6.2
1984	77	6.9
1985	78	8.0
1986	84	9.2
1987	84	10.6
1988	88	11.7
1989	88	12.3
1990	85	13.1
1991	84	13.7
1992	85	15.4
1993	87	17.8
1994	92	20.8
1995	92	24.4
1996	94	26.7
1997	94	28.7
1998	88	30.6
1999	94	33.4
2000	96	35.5
2001	93	37.8
2002	93	39.8

Source: FAO.

Energy and Climate Trends

SEGS III-VII Solar Power Facility at Kramer Junction, California

▶ Fossil Fuel Use Surges

▶ Nuclear Power Rises Once More

▶ Global Wind Growth Continues

▶ Solar Energy Markets Booming

▶ Biofuel Use Growing Rapidly

▶ Climate Change Indicators on the Rise

Fossil Fuel Use Surges

Christopher Flavin

World use of oil—the dominant fossil fuel—surged by 3.4 percent in 2004, the fastest rate of increase in 16 years, and reached an average of 3,760 million tons of oil equivalent.[1] (See Figure 1.) Oil producers had difficulty keeping up with soaring demand—estimated by the International Energy Agency at 82.4 million barrels per day—pushing prices to a record nominal level of $55 per barrel in October before retreating at the end of the year.[2]

Although precise figures were not yet available, use of natural gas and coal also appears to have surged in 2004.[3] (See Figure 2.) The continuing rapid growth in coal use in China and India, where pollution controls are minimal, is adding to local and long-distance pollution, ranging from sulfur and nitrogen oxides to mercury.[4]

pp. 40, 50, 88, 94

China and the United States were the main engines driving fossil fuel markets in 2004, accounting between them for nearly half the increase in world oil demand. China alone increased its oil consumption 11 percent in 2004, cementing its position as the world's number two user at 6.6 million barrels per day.[5] (See Figure 3.) The United States increased its oil use to 20.5 million barrels a day—nearly 25 percent of the world total.[6]

The jump in oil use in 2004 stemmed from a dramatic rebound in the world economy and the entrance of large sections of the developing world into oil-intensive stages of economic development. Not only are automobile numbers rising rapidly, but oil is popular in industry and power generation wherever the infrastructure to use gas and coal is stretched thin. In China, 24 of 31 provinces were subject to power rationing in 2004, pushing many factory owners to install diesel generators and driving up oil demand.[7]

Growth in world oil use is expected to slow in 2005 to a more normal 2 percent.[8] Still, oil prices early in the year were gyrating around $45 per barrel—roughly double the $20–30 that was typical in the 1990s.[9]

Analysts disagree on whether the higher oil prices are an aberration caused by temporary constraints such as terrorism in Saudi Arabia and Iraq or something more fundamental. Some believe that enough oil remains for world production to keep rising indefinitely—up 40 percent to 115 million barrels per day by 2020, according to the International Energy Agency.[10] But a growing number of geologists question whether oil reserves are sufficient to keep production rising. For the past three decades, they argue, oil companies have not been finding as much oil as they have been extracting—a gap that has widened in recent years.[11]

Oil production is already falling in 33 of the 48 largest-oil producing countries, including 6 of the 11 members of the Organization of the Petroleum Exporting Countries.[12] In the continental United States (excluding offshore), oil production peaked at 8 million barrels per day in 1970 and fell to just 2.9 million barrels a day in 2004.[13] During the past few years, Russia and the Persian Gulf countries have accounted for most of the increase in world production. Russia's oil industry is still rebounding from its post-Soviet collapse, but output began to level off by late 2004.[14]

In 2004, Saudi Arabia and the other Persian Gulf countries were producing near their historic peaks of the early 1980s, and for the first time in decades they were down to roughly a million barrels per day of spare capacity.[15] Although some new oil fields came on line in late 2004 and others are planned in 2007, some analysts doubt the region's ability to continually boost production.[16] Some of the largest oil fields in the Persian Gulf are more than 30 years old, and no independent verification of their claimed oil reserves has been permitted for decades.[17]

These developments suggest that the relatively stable oil prices of the 1990s are not likely to reappear anytime soon. PFC Energy, a Washington-based forecasting group that has carefully analyzed global reserve figures, concluded in 2004 that world oil production might be unable to meet projected demand as early as the middle of the next decade.[18] PFC projects that global production will peak in the next 10–15 years.[19] In a world accustomed to sustaining demand growth of roughly 2 percent a year, that would be a crude shock—one that would drive prices through the roof.

Figure 1. World Oil Consumption, 1950–2004

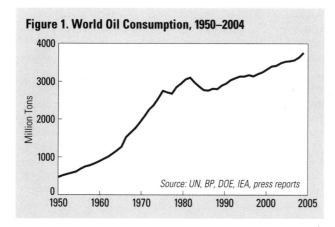

Source: UN, BP, DOE, IEA, press reports

Figure 2. World Consumption of Coal and Natural Gas, 1950–2003

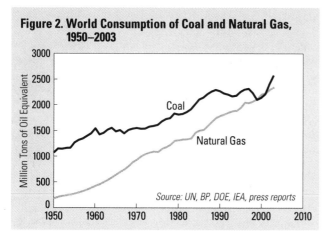

Source: UN, BP, DOE, IEA, press reports

Figure 3. Oil Consumption and Production in China, 1973–2004

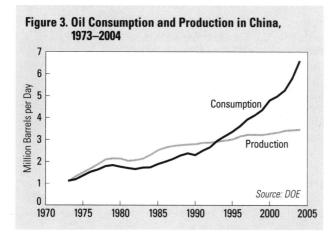

Source: DOE

World Fossil Fuel Consumption, 1950–2004

Year	Oil	Coal	Natural Gas
	(million tons of oil equivalent)		
1950	470	1,074	171
1955	694	1,270	266
1960	951	1,544	416
1965	1,530	1,486	632
1970	2,254	1,553	924
1971	2,377	1,538	988
1972	2,556	1,540	1,032
1973	2,754	1,579	1,059
1974	2,710	1,592	1,082
1975	2,678	1,613	1,075
1976	2,852	1,681	1,138
1977	2,944	1,726	1,169
1978	3,055	1,744	1,216
1979	3,103	1,834	1,295
1980	2,972	1,814	1,304
1981	2,868	1,826	1,318
1982	2,776	1,863	1,322
1983	2,761	1,916	1,340
1984	2,809	2,011	1,451
1985	2,801	2,107	1,493
1986	2,893	2,143	1,504
1987	2,949	2,211	1,583
1988	3,039	2,261	1,663
1989	3,088	2,293	1,738
1990	3,136	2,270	1,774
1991	3,134	2,225	1,806
1992	3,170	2,203	1,836
1993	3,139	2,168	1,869
1994	3,199	2,182	1,876
1995	3,246	2,255	1,937
1996	3,323	2,302	2,033
1997	3,398	2,315	2,024
1998	3,417	2,233	2,059
1999	3,485	2,103	2,106
2000	3,526	2,141	2,194
2001	3,538	2,211	2,217
2002	3,563	2,412	2,286
2003	3,637	2,578	2,332
2004 (prel)	3,760	n.a.	n.a.

Source: UN, BP, DOE, IEA, press reports.

Nuclear Power Rises Once More

Nicholas Lenssen

Between 2003 and 2004, total installed nuclear generating capacity increased by more than 2 percent, from 358,000 megawatts to nearly 366,000 megawatts.[1] (See Figure 1.) This figure, the highest ever reached, is roughly 8 percent greater than a decade ago, illustrating how modest nuclear energy's overall growth has been.[2] Yet with only 16,000 megawatts of new reactors currently under active construction, nuclear power is likely to grow at a slower pace in the next 10 years than it did in the last 10—or perhaps even shrink.[3]

The increase in 2004 came as six new reactors and one previously mothballed reactor were connected to the grid, accounting for 6,615 megawatts.[4] The remainder came from squeezing more power from existing reactors in countries such as the United States, France, and Spain.

New construction started on only one reactor in 2004, in India.[5] (See Figure 2.) Meanwhile, five reactors were permanently shut down, bringing the total to 114 (representing 33,663 megawatts).[6] (See Figure 3.)

In the United States, it has been more than 30 years since a new reactor order was placed and not subsequently cancelled, but government and industry efforts are still focused on a revival. Despite the failure of legislation that would have created major new subsidies, the government's Nuclear 2010 program took small steps toward early preconstruction licensing of new units.[7]

Efforts for new construction are further along in parts of Europe. In 2003, a Finnish utility consortium ordered two new reactors, the first new project in Europe in more than a decade.[8] And France announced that it would start building its first new reactor since the early 1990s, starting in 2007.[9]

Elsewhere in Europe, however, attention was focused on closing existing reactors and not building new ones. In the United Kingdom, four older, small reactors were permanently closed.[10] And Sweden reaffirmed its commitment to close a reactor in 2005 as part of the country's nuclear phaseout.[11]

Likewise, one of the Ignalina reactors in Lithuania was permanently shut down in 2004, with the second of the Chernobyl-style units to be closed in 2009.[12] Meanwhile, Romania abandoned its efforts to raise private financing to complete a reactor that construction started on initially in 1984, though work on a second reactor, initially started in 1982, is scheduled to be completed in 2007.[13]

One reactor was completed in Russia and two in Ukraine, all of which were started in the 1980s.[14] Work on new units at additional sites in Russia apparently was minimal due to lack of funding, as the industry focused on extending the life of operating reactors.[15] Still, five older reactors are to be shut by 2010.[16]

Japan's nuclear dreams continue to become nightmares due to accidents—the latest being a 2004 steam leak that killed four workers at a plant in Fukui Prefecture.[17] One new reactor was connected to the nation's grid, the first in roughly three years, leaving only three reactors under construction in Japan.[18]

South Korea completed one reactor in 2004, leaving just one unit being built.[19] The country plans to construct 10 new reactors despite growing public opposition.[20] The revelation that South Korean scientists worked on atomic weapons technologies as recently as 2000 also raised international concerns.[21] Indeed, a twin reactor project being built by the United States and allies in North Korea remains on hold following confirmation that the country continues to pursue nuclear weapons capabilities.[22]

China and India are the two countries with the most ambitious nuclear plans today. Despite having just nine reactors operating for 6,587 megawatts and two reactors under construction, China plans to boost its nuclear capacity to 36,000 megawatts by 2020—a daunting goal that would require the equivalent of finishing two new large reactors every year.[23] (Even then, however, nuclear would only represent 4 percent of China's total generating capacity.)[24]

India already has nine reactors under construction—nearly half of all reactors being built globally—but most are small capacity power plants, totaling just 4,122 megawatts in capacity.[25] As a result, the country's nuclear industry produces only 3.3 percent of India's electricity.[26]

Figure 1. World Electrical Generating Capacity of Nuclear Power Plants, 1960–2004

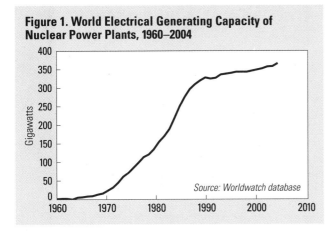

Source: Worldwatch database

Figure 2. World Nuclear Reactor Construction Starts, 1960–2004

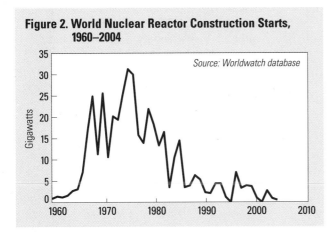

Source: Worldwatch database

Figure 3. Nuclear Capacity of Decommisioned Plants, 1964–2004

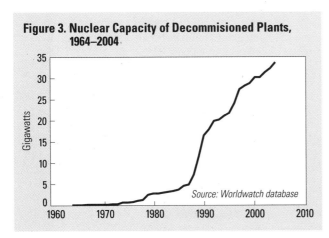

Source: Worldwatch database

World Net Installed Electrical Generating Capacity of Nuclear Power Plants, 1960–2004

Year	Capacity
	(gigawatts)
1960	1
1965	5
1970	16
1971	24
1972	32
1973	45
1974	61
1975	71
1976	85
1977	99
1978	114
1979	121
1980	135
1981	155
1982	170
1983	189
1984	219
1985	250
1986	276
1987	297
1988	310
1989	320
1990	328
1991	325
1992	327
1993	336
1994	338
1995	340
1996	343
1997	343
1998	343
1999	346
2000	349
2001	352
2002	357
2003	358
2004	366

Source: Worldwatch Institute database, IAEA, and press reports.

Global Wind Growth Continues

Janet L. Sawin

An estimated 8,210 megawatts of wind energy capacity were added globally in 2004, bringing the total to approximately 47,760 megawatts—enough to provide power to more than 22 million average homes in Europe.[1] (See Figure 1.) Since 2001, global wind capacity has nearly doubled.[2] Wind is the world's fastest-growing energy source after solar power, driven by falling costs, concerns about climate change, and strong government policies.

Annual installations reached a new record—8 percent higher than in 2003.[3] (See Figure 2.) Growth rates have fallen in recent years as onshore markets slow in some traditional powerhouses and offshore projects face slow starts. But several countries are positioned to become new leaders and could drive growth rates back up.

Wind turbines are operating in more than 65 countries—yet 72 percent of global capacity is spinning in Europe.[4] The European Union added 5,703 megawatts of capacity in 2004, bringing its total to 34,205 megawatts.[5]

For the first time, Spain surpassed the German market in 2004, adding 2,065 megawatts and becoming the world's top installer.[6] With 8,263 megawatts total, Spain meets 6 percent of its electricity demand with the wind.[7] The government has proposed raising national targets for 2011 from 13,000 to 20,000 megawatts.[8]

Germany remains the overall leader, with 16,629 megawatts of wind power, but added only 2,037 megawatts in 2004, the second consecutive year of market contraction.[9] The slowdown was due to uncertainty over renewable energy legislation and changing approval processes.[10] Yet wind energy now meets 6.6 percent of Germany's electricity needs, up from 3 percent in late 2001, and meets well over one fifth of power demand in four states.[11] Germany and Spain together now account for more than half of global wind capacity.[12]

The United States added only 389 megawatts in 2004, bringing the U.S. total to 6,740 megawatts—third place worldwide.[13] The U.S. market saw lackluster growth due to late extension of a federal tax credit, which has expired three times since it was first enacted a decade ago. Several projects were cancelled or frozen and more than 2,000 people lost work.[14] But with the credit in place for another year, the industry expects record additions in 2005.[15]

Also noteworthy were Italy and the Netherlands, which both passed the 1,000-megawatt mark in total installed capacity in 2004.[16] Despite having the best wind resources in Europe, the United Kingdom ended the year with only 888 megawatts.[17] But the U.K. market is picking up as well. More capacity was added in 2004 than in the three previous years combined.[18] Nearly 5,000 megawatts of onshore projects are in the planning stage, while planned offshore projects represent 7,200 megawatts—or 7 percent of national electricity demand.[19]

Denmark had a disappointing year, adding only 9 megawatts and barely hanging onto its fourth place position ahead of India.[20] But it still leads the world for offshore wind installations.[21] Offshore wind energy has developed slowly thus far, but studies show that such projects could meet a large share of Europe's electricity needs.[22]

Asia's wind energy market also picked up speed in 2004. It was led by India, which installed 875 megawatts for a total of 3,000 megawatts in place.[23] Wind power now accounts for 3 percent of India's electric capacity, and domestic manufacturing potential exceeds 1,000 megawatts per year.[24] China's capacity rose 35 percent, for a total of almost 770 megawatts.[25] At a major renewables conference in Germany in 2004, the Chinese government announced ambitious targets of 4,000 megawatts of wind energy by 2010, and five times that a decade later.[26]

Wind energy technology continues to advance. Today wind power is cheaper than natural gas even without subsidies. And on good sites, wind is closing in on coal.[27] The world's largest turbine—with 5 megawatts rated capacity—began operation in Germany in late 2004.[28] Global sales of wind power equipment topped $10 billion in 2004 and are projected to reach $49 billion a year by 2012.[29] The global wind industry now employs well over 100,000 people, and Germany alone expects to have more than 100,000 wind jobs by 2010.[30]

Figure 1. World Wind Energy Generating Capacity, 1980–2004

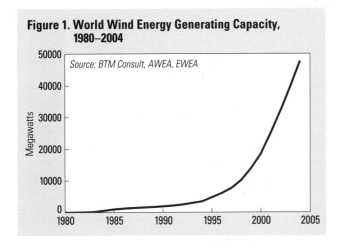

Source: BTM Consult, AWEA, EWEA

Figure 2. Annual Additions to World Wind Energy Generating Capacity, 1980–2004

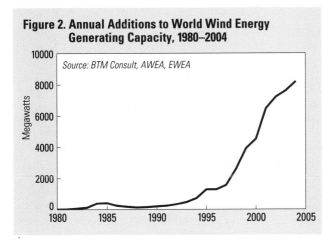

Source: BTM Consult, AWEA, EWEA

World Wind Energy Generating Capacity, Total and Annual Additions, 1980-2004

Year	Total	Annual Additions
	(megawatts)	
1980	10	5
1981	25	15
1982	90	65
1983	210	120
1984	600	390
1985	1,020	420
1986	1,270	250
1987	1,450	180
1988	1,580	130
1989	1,730	150
1990	1,930	200
1991	2,170	240
1992	2,510	340
1993	2,990	480
1994	3,490	730
1995	4,780	1,290
1996	6,070	1,290
1997	7,640	1,570
1998	10,150	2,600
1999	13,930	3,920
2000	18,450	4,520
2001	24,930	6,480
2002	32,037	7,227
2003	39,664	7,627
2004 (prel)	47,760	8,210

Source: BTM Consult, AWEA, EWEA.

Solar Energy Markets Booming

Janet L. Sawin

Global production of photovoltaic (PV) cells, which convert sunlight directly to electricity, reached an estimated 1,200 megawatts in 2004—a 58-percent jump over 2003 levels and a doubling of production in just two years.[1] (See Figure 1.) Strong policies in a handful of industrial countries have spurred production growth rates that average 43 percent a year since 2000.[2] Cumulative production, at 4,365 megawatts, has increased at an average annual rate of 32 percent since 2000, making PVs the world's fastest-growing energy source.[3]

An estimated 62 percent of new installations in 2004 were for grid-connected power, up from 3 percent a decade ago.[4] Solar power meets less than 1 percent of global electricity demand, but this threshold will soon be crossed if rapid growth continues.[5] The industry supports more than 25,000 jobs worldwide, and PVs are now a $7-billion market.[6] Analysts expect sales to reach $30 billion by 2010.[7]

Japan is the world leader, accounting for over 50 percent of PV production in 2004.[8] (See Figure 2.) A six-year government initiative encouraged the installation of 1,000 megawatts of capacity, and at least 160,000 Japanese homes are now PV-powered.[9] The government has set a target of 4,820 megawatts installed by 2010 and aims for PVs to generate 10 percent of Japan's electricity by 2030.[10]

Driven by strong government incentive programs, Europe produced 27 percent of new solar cells in 2004 and passed Japan in annual installations.[11] PV capacity in Germany surged by some 300 megawatts, more than any other country and double the preceding year's installations, bringing the nation's total to about 700 megawatts.[12] Three large Bavarian solar parks alone added 10 megawatts, enough to power 9,000 German homes.[13]

The United States remains a major producer, but its share of the global market has declined steadily—from 44 percent in 1996 to 11 percent in 2004.[14] Yet new installations continue to rise, and cumulative capacity reached about 277 megawatts by late 2003.[15] Thanks to supportive state policies, California leads the nation in PV use, with more than 93 megawatts installed.[16]

Technological advances, scale economies in production, and experience installing systems have led to significant cost reductions. In Japan, the average price for residential PV systems has declined more than 80 percent since 1993.[17] Globally, module costs have dropped from about $30 per watt in 1975 to close to $3 per watt.[18] Costs increased slightly in 2004 as demand outpaced supply and as silicon prices rose.[19] Still, PVs are the cheapest option for many remote or off-grid functions, and even on-grid they are competitive during peak demand in California and at all times in Japan.[20] Advanced technologies under development could cut costs further and revolutionize the power industry.[21]

The market for solar thermal collectors, which capture the sun's warmth to heat water and building space, is also booming. The global market grew some 50 percent between 2001 and 2004.[22] (See Figure 3.) About 18 million square meters of capacity were added in 2004, bringing global installations for all uses, including swimming pools, to an estimated 150 million square meters.[23] The energy equivalent of this capacity far exceeds that of global wind and solar power combined.[24] About 73 percent of this total heats water and space, meeting the needs of more than 32 million households worldwide; the rest is used for pools.[25]

China, long the world's leader in solar thermal production and use, accounted for 55 percent of global solar heating capacity (excluding pool systems) by the end of 2003, or 52 million square meters of collectors.[26] Government targets call for more than four times that area by 2015.[27] Other large markets include Japan, Europe, and Turkey, each with at least 10 percent of global capacity.[28]

The United States, once the world leader, is now far behind due to low natural gas prices and the elimination of incentives after the 1980s. Today 98 percent of U.S. solar systems heat pools, a market that hardly exists elsewhere.[29] Yet the solar heating industry has the potential for spectacular energy savings and market growth—even in the United States, where home solar water heating systems pay for themselves in four to eight years through fuel savings.[30]

Figure 1. World Photovoltaic Production, 1971–2004

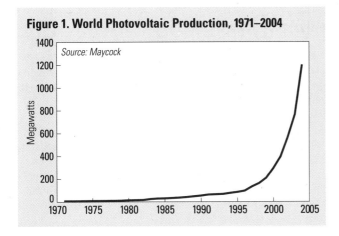

Figure 2. Photovoltaic Production by Country or Region, 1994–2004

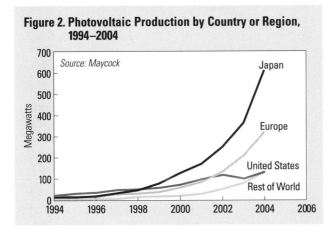

Figure 3. Global Solar Water Heating Annual Installations, Excluding Pool Systems, 1998–2004

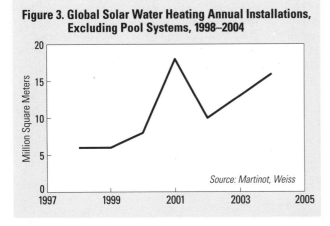

World Photovoltaic Production, 1971–2004

Year	Annual Production	Cumulative Production
	(megawatts)	
1971	0.1	0.1
1975	1.8	1.9
1980	7	19
1985	23	98
1990	47	273
1991	55	329
1992	58	387
1993	60	447
1994	69	516
1995	78	594
1996	89	682
1997	126	808
1998	155	963
1999	201	1,164
2000	288	1,452
2001	391	1,842
2002	562	2,404
2003	761	3,165
2004 (prel)	1,200	4,365

Source: Maycock.

Global Solar Water Heating Installations, Excluding Pool Systems, 1997–2004

Year	Annual Installations	Cumulative Installations
	(million square meters)	
1997	n.a.	33
1998	6	39
1999	6	45
2000	8	53
2001	18	71
2002	10	81
2003	13	94
2004 (prel)	16	110

Source: Martinot, Weiss.

Biofuel Use Growing Rapidly

Molly Aeck

Production and use of biofuels—fuels derived from crops and agricultural wastes—advanced rapidly in 2004, spurred on by agricultural, environmental, and consumer interests. In general, biofuels burn cleaner than fossil fuels, are renewable, and can be domestically produced in many countries—creating agricultural jobs and revenues while displacing imported fuels.

Global production of fuel ethanol increased 13.6 percent in 2004, reaching almost 33 billion liters.[1] (See Figure 1.) Nearly twice as much ethanol was produced in 2004 as in 2000.[2] Ethanol is by far the most widely used biofuel for transportation; Brazil and the United States dominate the market. World production of biodiesel fuel, based on vegetable oils and fats, is smaller but has been growing even faster, nearing 1.8 billion liters in 2003, up 18 percent over 2002.[3] (See Figure 2.)

LINKS p. 56

Ethanol derived from sugarcane accounts for 30 percent of auto fuel in Brazil, which generates some 14 billion liters of ethanol annually.[4] In the United States, corn-distilled ethanol provides more than 10 billion liters of fuel each year, but this accounts for just 2 percent of U.S. transportation fuel.[5]

The ethanol fuels market grew rapidly in the 1980s due to Brazilian and U.S. government efforts to provide alternatives to high-priced oil, but then it languished for much of the 1990s. Since 2000, however, rising environmental concerns, new technologies, and the desire to find new income streams for farmers have provided a large boost.

The European Union (EU) is the third largest producer of biofuels but the leading manufacturer of biodiesel. With the help of tax breaks for diesel fuel, nearly 1.6 billion liters of biodiesel were produced in Europe in 2003, a 43-percent increase over 2001.[6] While conventional diesel vehicles can run on 5–30 percent blends of bio- and fossil diesel, several European vehicle manufacturers have approved the use of 100 percent biodiesel in their engines.[7] The EU hopes biofuels will supply 2 percent of the fuel market in 2005, 5.75 percent in 2010, and 20 percent in 2020.[8]

The growth of biofuels may accelerate even more as others introduce favorable policies. Australia, China, India, South Korea, and Japan already support biofuels.[9] The government of Thailand has endorsed a 10-percent ethanol/gasoline blend, and 18 new ethanol plants are being developed.[10] In the Philippines, coconut-derived biodiesel is expected to cut demand for petroleum diesel by 5 percent.[11]

The cost of biofuels varies widely by region. In Brazil, for example, the retail price of ethanol is often lower than that of gasoline due to low land and labor costs. In North America, in contrast, ethanol is more expensive because of the lower efficiency of corn as opposed to sugarcane and higher costs.

The greatest potential for biofuels lies in tropical and subtropical developing countries, where growing seasons are longer and production costs are lower.[12] But unlike oil, trade in biofuels is limited by tariffs and other trade restrictions.[13] Although producing ethanol costs about half as much in Brazil as in Europe, ethanol trade between the two is nearly nonexistent. In 2004, Brazil exported 2.3 billion liters of ethanol to India, the United States, and the Caribbean, a figure that could be much higher if Brazil were not constrained by high agricultural tariffs.[14]

Reducing the cost of biofuels is the key to their continued growth. New conversion technologies, such as cellulose-derived ethanol made from the non-food portion of renewable feedstocks, could bring significant cost reductions over the next decade. Canada-based Iogen, the world leader in cellulose ethanol technology, now produces approximately 100,000 liters a year; several new plants are planned that could quadruple the country's ethanol supply.[15] Less expensive processes for biodiesel production are also being developed.

The International Energy Agency projects that if supportive policies continue to proliferate, world biofuels production could nearly quadruple, to more than 120 billion liters, by 2020.[16] More than 2 million additional alternative fuel vehicles could be introduced worldwide by 2010, driving up demand for biofuels.[17]

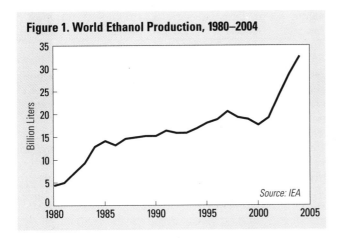

Figure 1. World Ethanol Production, 1980–2004

Source: IEA

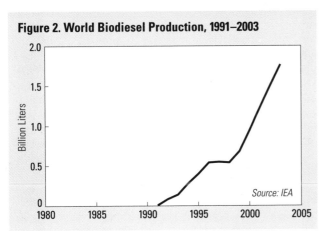

Figure 2. World Biodiesel Production, 1991–2003

Source: IEA

World Ethanol and Biodiesel Production, 1980–2004

Year	Ethanol	Biodiesel
	(million liters)	
1980	4,368	
1981	4,977	
1982	7,149	
1983	9,280	
1984	12,880	
1985	14,129	
1986	13,193	
1987	14,599	
1988	14,902	
1989	15,191	
1990	15,190	
1991	16,348	11
1992	15,850	88
1993	15,850	143
1994	16,829	283
1995	18,033	402
1996	18,789	542
1997	20,562	550
1998	19,247	542
1999	18,840	683
2000	17,580	949
2001	19,136	1,231
2002	24,106	1,504
2003	28,745	1,768
2004	32,655	n.a.

Source: International Energy Agency.

Climate Change Indicators on the Rise

Janet L. Sawin

In 2004, the average atmospheric carbon dioxide (CO_2) concentration reached 377.4 parts per million by volume.[1] (See Figure 1.) The average CO_2 concentration has increased more than 19 percent since measurements began at Mauna Lao Observatory in Hawaii in 1959—and has gone up 35 percent since the dawn of the industrial age.[2] Average annual rates of increase have more than doubled since 1960.[3]

The average global temperature actually fell slightly in 2004, but at 14.48 degrees Celsius the year was still the fourth warmest since 1880, according to the Goddard Institute for Space Studies.[4] (See Figure 2.)

LINKS pp. 30, 56, 60, 88

Other climate analysis centers, using roughly the same network of land- and sea-based weather stations, also rank 2004 behind only 1998, 2002, and 2003.[5] Since the early 1900s, average global temperature has risen 0.6 degrees Celsius, but the rate of change since 1976 has been triple that for the century as a whole.[6] The 10 warmest years on record have all occurred since 1990.[7]

The impacts of rising CO_2 concentrations and temperatures are already visible worldwide and are arriving faster than feared, according to some climate experts.[8] The World Health Organization estimates that at least 160,000 people die annually due to climate change, and there is growing evidence of direct links to observed ecological changes.[9]

Higher temperatures and precipitation changes have driven species northward or to higher elevations and have affected the timing of breeding and migratory seasons. Carbon cycling and storage processes have been altered. Mountain glaciers are shrinking at ever-faster rates, threatening water supplies for millions of people and species.[10] A study by the U.S. National Center for Atmospheric Research found that rising global temperatures have been a key factor in increasing drought worldwide.[11]

The effects are most pronounced in the Arctic, where in recent decades temperatures have risen at almost twice the average rate of the rest of the world.[12] The average area of summer sea ice cover in the region has declined by 15–20 percent over the past 30 years, shrinking habitat for polar bears, caribou, and other Arctic species, while sea level there has risen 10–20 centimeters over the past century.[13]

Preliminary data indicate that fossil fuel burning released more than 7 billion tons of carbon in 2004, an increase of at least 3 percent over 2003, continuing the accelerating release rate of that year, when emissions rose 3.8 percent.[14] (See Figure 3.) Carbon emissions from fossil fuels are believed to be the main factor behind the rise in atmospheric concentrations and global temperatures.[15] Nearly three times as much carbon was released in 2004 as in 1960.[16]

Ten countries are responsible for about two thirds of global carbon emissions from fuel use. The United States, with 5 percent of the world's population, accounts for nearly a quarter of the total.[17] Between 1990 and 2003, U.S. energy-related emissions rose 16 percent.[18] China ranks second, with a 14-percent share.[19] Emissions there are up more than 47 percent since 1990, and China accounted for half of the global increase in 2003, although it still ranks far behind the industrial world in emissions per person.[20]

Although global emissions and temperatures continue their upward climb, some progress on slowing climate change has been made. Russia ratified the Kyoto Protocol in October 2004, enabling the treaty to enter into force in February 2005.[21] Although the agreement is widely seen as inadequate to address the rising threat, it takes the first steps toward that goal.

At the same time, several countries have adopted strong policies to promote a shift from fossil fuels to renewable energy technologies, including Germany, Spain, Japan, and the Philippines. China is in the process of drafting a major renewable energy law and has set ambitious targets that call for renewables to meet 17 percent of China's projected energy consumption by 2020.[22]

The world's first international emissions trading scheme began operating in the European Union on 1 January 2005.[23] Some analysts believe that the carbon business could become one of the world's largest commodity markets.[24]

Figure 1. Atmospheric Concentrations of Carbon Dioxide, 1960–2004

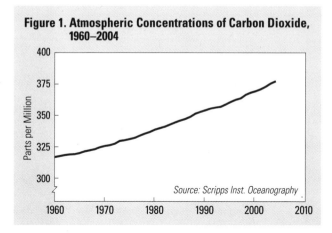

Source: Scripps Inst. Oceanography

Figure 2. Global Average Land-Ocean Temperature at Earth's Surface, 1880–2004

Source: GISS

Figure 3. Carbon Emissions from Fossil Fuel Burning, 1950–2003

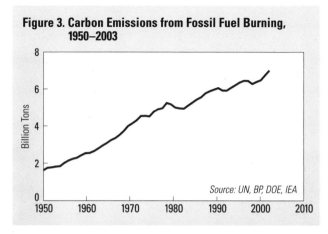

Source: UN, BP, DOE, IEA

Global Average Temperature and Carbon Emissions from Fossil Fuel Burning, 1950–2004, and Atmospheric Concentrations of Carbon Dioxide, 1960–2004

Year	Carbon Dioxide (parts per mill. by vol.)	Temper- ature (degrees Celsius)	Emissions (mill. tons of carbon)
1950	n.a.	13.87	1,612
1955	n.a.	13.89	2,013
1960	316.9	14.01	2,535
1965	320.0	13.90	3,087
1970	325.7	14.02	3,997
1975	331.2	13.94	4,518
1980	338.7	14.16	5,177
1981	339.9	14.22	5,004
1982	341.1	14.07	4,959
1983	342.8	14.25	4,942
1984	344.4	14.07	5,113
1985	345.9	14.04	5,274
1986	347.2	14.12	5,436
1987	348.9	14.27	5,559
1988	351.5	14.30	5,774
1989	352.9	14.19	5,881
1990	354.2	14.37	5,969
1991	355.6	14.32	6,053
1992	356.4	14.14	5,921
1993	357.0	14.14	5,917
1994	358.9	14.25	6,067
1995	360.9	14.38	6,205
1996	362.6	14.24	6,350
1997	363.8	14.40	6,445
1998	366.6	14.56	6,440
1999	368.3	14.33	6,274
2000	369.5	14.31	6,385
2001	371.0	14.47	6,479
2002	373.1	14.54	6,743
2003	375.6	14.52	6,999
2004(prel)	377.4	14.48	7,210

Source: GISS, BP, IEA, CDIAC, DOE, and Scripps Inst. of Oceanography.

Economic Trends

Chemonics/M. Herrick, courtesy of USAID

Worker in a Ugandan nursery that grows flowers for export

▶ Global Economy Continues to Grow

▶ World Trade Rises Sharply

▶ Foreign Direct Investment Inflows Decline

▶ Weather-Related Disasters Near a Record

▶ Steel Surging

Global Economy Continues to Grow

Erik Assadourian

Gross world product (GWP)—the aggregated estimate of total output of goods and services in countries around the world—increased 5 percent in 2004, to $55 trillion (in 2003 dollars).[1] (See Figure 1.) This rapid growth was primarily driven by expansion in industrial markets and by explosive growth in emerging markets, particularly China.[2] Yet with world population increasing by 73 million in 2004, per capita GWP grew less rapidly, rising 3.8 percent to $8,587.[3]

The U.S. gross domestic product (GDP) grew 4.3 percent in 2004, driven by domestic consumption and business investment, though high energy prices curbed growth late in the year.[4] Japan also demonstrated strong growth at 4.4 percent, propelled by business investments, exports, and a resurgence in domestic demand.[5] The European Union's economy expanded too, though more slowly—with GDP increasing by 2.2 percent.[6] Much of this growth came from exports.[7]

LINKS pp. 46, 52, 56, 108

Some of the most impressive expansion occurred in Asian developing countries, particularly China and India, which grew at 9.0 percent and 6.4 percent respectively.[8] Both benefited from significant foreign investment, increasing domestic demand, and a recovery in the information technology sector.[9] Africa's economy grew 4.5 percent, driven primarily by improved access to industrial-country markets, reduced debt burdens, and high commodity prices, particularly oil.[10] The Middle East's economy also benefited from high oil prices, growing 5.1 percent.[11] With the region's oil production now nearing capacity, however, economic growth is plateauing.[12]

In recent years, an increasing array of experts, institutions, and even governments have questioned the value of GDP as an accurate measure of economic growth or national economic progress. The primary failing is that GDP is an absolute measure. Thus all expenditures—regardless of their worth to society—are counted as positives.[13] Moreover, the worth of some essential economic sectors, like subsistence farming and household maintenance, is completely omitted.[14]

Another flaw is GDP's omission of economic externalities, like resource depletion and pollution. As human economic systems depend on natural resources and services, such as waste treatment and climate regulation, the failure to incorporate these into economic measures minimizes the worth of these ecosystem services. One analysis of humanity's consumption of renewable resources finds that humanity is using resources 21 percent faster than Earth can renew them.[15] (See Figure 2.) This conservative estimate, which does not include the needs of other species, nonrenewable resource use, or pollution, notes that on average each person uses the resources of 2.2 "global hectares" of productive land.[16] Yet only 1.8 global hectares on average is available per person worldwide.[17]

To counter the failings of the GDP measure, Redefining Progress, a U.S. nongovernmental research group, created the genuine progress indicator (GPI). This alternative measure adds ignored sectors like unpaid child care and volunteer work, while subtracting uncounted economic costs such as traffic, pollution, and crime. In the United States, per capita GDP grew 56 percent from 1982 to 2002.[18] Yet per capita GPI grew just 2 percent during that period, because the added value of beneficial services was almost entirely countered by growth in pollution and other social ills.[19] (See Figure 3.)

While the Redefining Progress initiative has drawn attention to the flaws of GDP, most promising is the Chinese government's plan to start incorporating environmental costs into its economic calculations. In 2004, China announced that it would implement a "Green GDP" measure in the next five years that would subtract resource depletion and pollution costs from GDP.[20] Early research suggests that China's average GDP growth between 1985 and 2000 would have been 1.2 percent lower had environmental costs been subtracted.[21] If fully implemented, not only would the Green GDP indicator help put China on a more sustainable economic path, it could push other major economies to follow suit—which in turn could transform the types of economic development the world values.

Figure 1. Gross World Product, 1950–2004

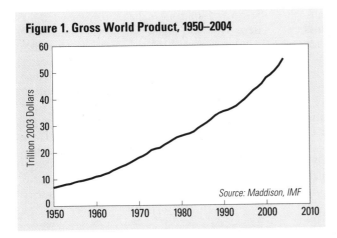

Source: Maddison, IMF

Figure 2. World Ecological Footprint, 1961–2001

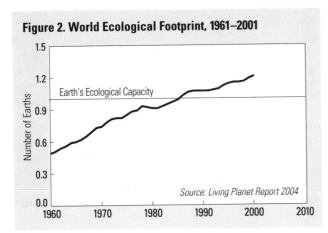

Earth's Ecological Capacity

Source: Living Planet Report 2004

Figure 3. GDP and GPI, United States, 1950–2002

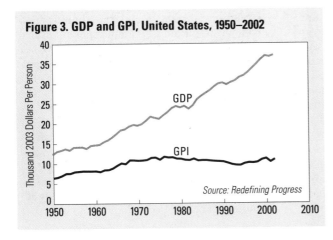

GDP

GPI

Source: Redefining Progress

Gross World Product, 1950–2004

Year	Total	Per Capita
	(trill. 2003 dollars)	(2003 dollars)
1950	6.9	2,710
1955	8.9	3,192
1960	11.0	3,607
1965	14.0	4,179
1970	17.9	4,825
1971	18.6	4,926
1972	19.5	5,057
1973	20.8	5,286
1974	21.3	5,308
1975	21.6	5,292
1976	22.7	5,454
1977	23.6	5,580
1978	24.6	5,729
1979	25.5	5,832
1980	26.0	5,849
1981	26.5	5,861
1982	26.9	5,827
1983	27.6	5,891
1984	28.9	6,056
1985	29.9	6,160
1986	30.9	6,270
1987	32.1	6,387
1988	33.5	6,551
1989	34.5	6,649
1990	35.2	6,671
1991	35.6	6,641
1992	36.3	6,668
1993	37.1	6,711
1994	38.4	6,843
1995	39.7	6,978
1996	41.3	7,149
1997	42.9	7,330
1998	44.0	7,415
1999	45.4	7,566
2000	47.6	7,823
2001	48.7	7,914
2002	50.2	8,056
2003	52.1	8,273
2004 (prel)	54.7	8,587

Source: Organisation for Economic Co-operation and Development and International Monetary Fund.

World Trade Rises Sharply

Zoë Chafe

The total value of world exports, a measure of the global trade linkages between countries, reached $9.2 trillion in 2003, according to estimates by the International Monetary Fund (IMF).[1] (See Figure 1.) This was the value of all products (such as food or building materials) and services (such as tourism) sold by people in one country to people in another that year. The IMF expected the value of world exports to reach $10.6 trillion in 2004, an increase of 15.3 percent over 2003.[2] This would be the highest growth rate since 1995, when the value rose 16.7 percent.[3]

LINKS pp. 44, 46, 52

The share of world exports in gross world product, which charts the proportion of the world's products that is exported each year, grew by 9.6 percent in 2004, reaching 17.7 percent—the highest level since 1997.[4] (See Figure 2.) This growth occurred despite a rapid rise in oil prices, which can depress both trade and output.[5] The average oil spot price reached $38.59 per barrel (in 2003 dollars), up from $31.06 in 2003.[6] High oil prices often affect oil-importing developing countries most severely, as they use on average twice as much oil to produce a unit of economic output as industrial countries do, and they may be less able to shoulder the additional financial burden of higher oil prices.[7]

Trade in oil, steel, and minerals expanded in 2004, influenced by growth in Chinese construction and manufacturing sectors.[8] China represented more than 20 percent of the increase in world trade volumes during 2004, and its share in world exports nearly doubled over the preceding four years, rising from 2.8 percent to 5.8 percent.[9] Its performance continues to be fueled by its relatively recent accession to the World Trade Organization (WTO) as well as by rapid rates of investment and consumption.[10]

The volume of world soybean exports during 2004/05 was expected to rise 16 percent over the previous year, reaching a total of 65 million tons.[11] The increase will be mainly in response to growth in soybean plantings spurred by a spike in soybean prices between mid-2003 and early 2004. This occurred because of droughts in the United States and Brazil—the two largest producers—combined with increasing demand from China, now the largest importer. China is expected to import 23 million tons of soybeans in 2004/05, more than twice as much as in 2001/02.[12] Most world market prices for agricultural commodities have rebounded over the past two years, after a downward trend during the late 1990s and 2001.[13]

Representatives from WTO member countries are now working to complete the Doha Round of trade negotiations following meetings in Doha, Qatar, in 2001. These negotiations are designed to address the Doha Development Agenda, which seeks to bolster developing countries' access to consumer markets in industrial nations by reducing trade barriers, especially in agriculture.

When the fourth ministerial conference of the WTO met in Cancun, Mexico, in August 2003, negotiations aimed at resolving the Doha Round fell through, forcing the summit to end with no consensus on a final document.[14] A key factor in the meltdown was that, in an unprecedented show of unity, groups of developing countries formed a cohesive voting bloc, refusing to agree to the draft text proposed by various industrial nations.[15]

Negotiations over the following year resulted in adoption of a revised agenda that rescued the Doha Round.[16] It mandates WTO members to consider contentious issues such as increasing the competitiveness of cotton exports by African farmers in a market of low prices for industrial-nation cotton exports, cutting agricultural export subsidies, and revamping rules on "special and differential treatment," which give developing-country members more time to reduce their agricultural subsidies.[17]

A recent report from the U.N. Food and Agriculture Organization echoed the need for WTO negotiations to give priority to reducing agricultural tariffs and export subsidies, warning that low agricultural commodity prices, often caused by market-distorting tariffs and subsidies, threaten the food security of hundreds of millions of people in developing countries.[18]

Figure 1. World Exports of Goods, 1950–2003, and of Goods and Services, 1970–2003

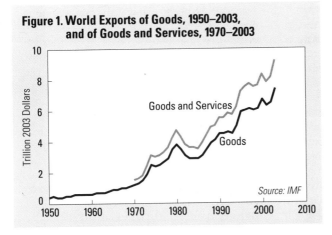

Source: IMF

Figure 2. World Exports of Goods and Sevices as a Share of Gross World Product, 1970–2003

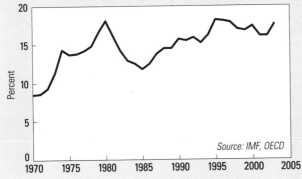

Source: IMF, OECD

World Exports of Goods and Services, 1950–2003

Year	Goods	Goods and Services
	(trillion 2003 dollars)	
1950	0.4	
1955	0.5	
1960	0.6	
1965	0.9	
1970	1.2	1.5
1971	1.3	1.6
1972	1.5	1.8
1973	1.9	2.4
1974	2.5	3.1
1975	2.4	3.0
1976	2.5	3.1
1977	2.7	3.3
1978	2.9	3.6
1979	3.5	4.2
1980	3.8	4.7
1981	3.5	4.3
1982	3.1	3.8
1983	2.9	3.6
1984	2.9	3.6
1985	2.9	3.5
1986	3.1	3.9
1987	3.5	4.4
1988	3.9	4.9
1989	4.1	5.0
1990	4.5	5.5
1991	4.5	5.5
1992	4.6	5.8
1993	4.5	5.7
1994	5.0	6.2
1995	5.9	7.2
1996	6.0	7.5
1997	6.1	7.7
1998	6.0	7.5
1999	6.1	7.6
2000	6.7	8.3
2001	6.3	7.8
2002	6.5	8.1
2003	7.4	9.2

Source: IMF.

Foreign Direct Investment Inflows Decline   Erik Assadourian

Foreign direct investment (FDI)—investment in enterprises from abroad that gives investors influence over the management of these enterprises—was the largest source of foreign capital in 2003, playing a significant role in shaping the global economy.[1] Yet flows of FDI to recipient countries fell 19 percent, to $560 billion.[2] These "inflows" have been declining since a peak of $1.47 trillion in 2000.[3] Preliminary data for 2004 suggest that this decline has ended, however, with FDI inflows projected to increase to $601 billion.[4] (See Figure 1.)

Reduced inflows to industrial countries were responsible for the whole decline in 2003, with their FDI falling 27 percent to $367 billion.[5] (See Figure 2.) Inflows to low- and middle-income countries, on the other hand, increased marginally to $193 billion.[6] In 2004, this trend appears to have continued, with inflows to industrial countries falling to $315 billion but those to low- and middle-income countries jumping to $286 billion.[7]

LINKS pp. 46, 98

In 2003, the United States was one of the hardest hit industrial countries, receiving 53 percent less in FDI than in 2002 (a total of $30 billion) and less than a tenth of what it got in 2000.[8] France became the largest industrial-country FDI recipient, at $47 billion, although it too experienced a minor dip in inflows (6 percent).[9]

Of low- and middle-income countries, the biggest recipient was China, absorbing $54 billion—on a par with the 2002 investments.[10] (When comparing inflows to the size of the respective economy, however, China ranked thirty-seventh in the world.)[11] Hong Kong, Singapore, Mexico, and Brazil were the next largest, together receiving $46 billion in 2003.[12] These five economies accounted for just over half of the inflows to low- and middle-income countries.[13] All of Africa, in contrast, received just $15 billion, though this did represent a 25-percent increase over 2002.[14]

The vast majority of FDI outflows originate from industrial countries. In 2003, 55 percent came from the European Union and 25 percent came from the United States.[15] Low- and middle-income countries account for only 7 percent.[16]

At $193 billion, FDI was the largest component of external capital flows in low- and middle-income countries in 2003.[17] Official development assistance (ODA), at $69 billion, also played a significant role, especially to many of the least developed countries, whose FDI inflows were relatively minor.[18] Yet while total ODA has hovered at the current level for the past decade, FDI inflows have increased 43 percent.[19]

Cross-border mergers and acquisitions (M&As) are one of the largest sources of FDI.[20] In 2003, these accounted for $297 billion, though the total value of cross-border M&As was down 20 percent, from $370 billion in 2002.[21]

The sectors receiving FDI have changed over the years. While investment directed toward primary industries such as agriculture and mining declined from 9 percent of total FDI in 1990 to 6 percent in 2002, and that in manufacturing dropped from 42 to 34 percent, the share going to services jumped from 49 to 60 percent.[22] In the primary sector, mineral and petroleum extraction accounted for the overwhelming majority of FDI inflows in 2001–02.[23] FDI inflows in the services sectors were more equally distributed, although finance and business services absorbed the largest shares.[24]

The benefits and costs of FDI inflows continue to be mixed. FDI can stimulate economic growth.[25] It can also lead to technology transfers, which can help improve efficiency and reduce environmental impacts.[26] But if profits from these investments are not reinvested or if interest payments on intracompany loans (a source of FDI) are overly burdensome, the economic benefits can be limited.[27] Further, if investments divert production away from traditional sectors and toward goods and services that are polluting or that stimulate unhealthy or unsustainable lifestyles, then FDI can have overall negative impacts.[28]

Whether a country benefits from FDI depends largely on the regulatory environment in the host country.[29] Without effective policies, for example, FDI can push local enterprises out of business or stimulate an inequitable distribution of services.[30]

Figure 1. Inflows of Foreign Direct Investment, 1970–2004

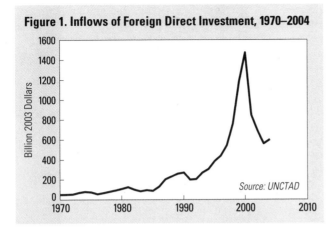

Source: UNCTAD

Figure 2. Inflows of FDI to Industrial and Low- and Middle-Income Countries, 1970–2004

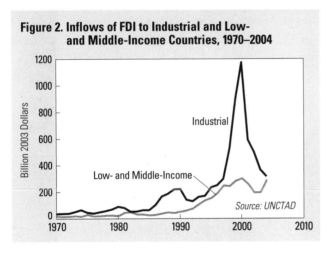

Industrial

Low- and Middle-Income

Source: UNCTAD

Inflows of Foreign Direct Investment, 1970–2004

Year	Total	Industrial Countries	Low- and Middle- Income Countries
		(billion 2003 dollars)	
1970	50	37	14
1971	51	39	12
1972	53	40	13
1973	69	52	17
1974	79	66	13
1975	74	47	27
1976	56	41	15
1977	67	50	17
1978	80	59	21
1979	93	71	22
1980	108	91	17
1981	125	82	42
1982	100	54	46
1983	84	55	29
1984	95	65	29
1985	88	65	23
1986	130	104	26
1987	203	167	36
1988	231	187	44
1989	260	220	40
1990	271	222	49
1991	199	142	58
1992	204	131	73
1993	270	164	106
1994	306	171	135
1995	386	235	151
1996	439	251	188
1997	542	298	244
1998	759	519	240
1999	1,177	897	280
2000	1,471	1,174	297
2001	846	592	255
2002	691	499	192
2003	560	367	193
2004 (prel)	601	315	286

Source: UNCTAD.

Weather-Related Disasters Near a Record

Molly Aeck

In 2004, weather-related natural disasters caused nearly $105 billion in economic losses (in 2003 dollars)—almost twice the total in 2003.[1] (See Figure 1.) This was only the second time that recorded losses from weather disasters have topped $100 billion (in constant dollars).[2] While the number of such events was below average for this decade, at 556, over the last 25 years there has been a general increase in disasters that are connected to weather, a category that includes storms, tornados, floods, heat waves, and extreme cold waves.[3] (See Figure 2.)

Some 12,000 people lost their lives in these weather disasters.[4] The fatality total would be far higher if all natural disasters were included in these data—particularly, of course, the December 2004 earthquake off Sumatra and the subsequent tsunamis it unleashed across the Indian Ocean. More than 280,000 people in 11 countries died in a matter of hours or days from these devastating events, and millions were left homeless.[5]

In summer 2004, surging floodwaters from annual monsoons killed more than 2,200 people in India, Nepal, and Bangladesh.[6] Hundreds of residents of Dhaka, Bangladesh, became sick after waters submerged 40 percent of the city.[7] Summer floods also affected China and the Caribbean, where regional death tolls neared 2,000.[8] In China's Yangtze River basin—where 85 percent of the original forest cover has been clearcut—flooding displaced 14 million people and destroyed 4.5 million hectares of cropland.[9] Heavy rainfall also surged down barren slopes in Haiti, where local forests have been stripped for fuelwood.[10]

Worldwide, the average annual occurrence of severe flooding has nearly doubled since the 1980s, from 110 to 205 events.[11] Scientists attribute the rise to deforestation and subsequent erosion, to irrigation that interferes with river drainage, and to global climate change, which may be increasing snowmelt and the intensity of storms and other extreme weather events.[12] Population growth and rising concentrations of people and wealth in cities and vulnerable areas are also contributing to increased economic and human losses from weather-related disasters.[13]

A string of hurricanes that hit the southern United States and the Caribbean made 2004 the costliest hurricane season ever. Damage from Hurricane Charley alone topped $20 billion.[14] Indeed, the insurance industry had its most expensive year for weather-related natural disaster payouts, covering nearly $42 billion of the estimated $105 billion in losses.[15] (See Figure 3.) Some 88 percent of these insured losses were linked to the U.S. and Caribbean hurricanes and to cyclones in Japan, which caused landslides, heavy structural damage, and loss of basic services.[16]

Roughly 12,000 weather-related disasters since 1980 have caused just over 618,200 fatalities and cost a total of $1.3 trillion.[17] Average annual economic losses from such events have risen from $26 billion in the 1980s to $67 billion over the last decade.[18] Average annual fatalities due to weather, meanwhile, jumped from 22,000 in the 1980s to 33,000 in the 1990s.[19]

Environmental disasters, including severe weather events, are also to blame for a large number of the world's refugees—30 million people, by one estimate.[20] Klaus Töpfer of the U.N. Environment Programme believes that the number of environmental refugees worldwide could reach 50 million by 2010.[21]

Long-term environmental management is key to reducing vulnerability to natural disasters. Many countries, however, face political instabilities that make such preventive measures unlikely, or they lack alternatives to poor resource management practices, such as the unsustainable use of wood for fuel. In many cases, the impact of weather and other natural disasters is mitigated by short-term solutions such as the construction of dikes and false banks, which provide only superficial security.[22]

Advances in the science and technology of hazard mitigation now provide a way to reduce losses significantly. The global scope of disasters requires that mitigation efforts be coordinated internationally. Public education and rapid communication networks are needed to transmit information on potential disaster risks in order to save lives and minimize property damage.

Figure 1. Economic Losses from Weather-Related Disasters, 1980–2004

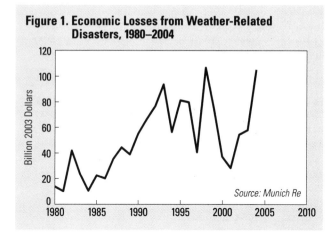

Source: Munich Re

Figure 2. Number of Weather-Related Disasters, 1980–2004

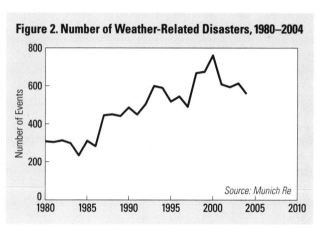

Source: Munich Re

Figure 3. Insured Losses from Weather-Related Disasters, 1980–2004

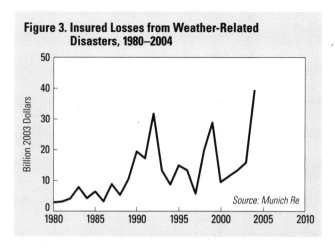

Source: Munich Re

Economic and Insured Losses from Weather-Related Disasters, 1980–2004

Year	Economic Loss	Insured Loss
	(billion 2003 dollars)	
1980	13.8	2.3
1981	10.1	2.5
1982	41.9	3.5
1983	23.5	7.4
1984	10.4	3.6
1985	22.5	5.8
1986	20.1	2.6
1987	35.4	8.3
1988	44.4	4.7
1989	39.0	10.0
1990	55.2	19.0
1991	66.3	16.8
1992	76.4	31.5
1993	93.6	12.8
1994	56.4	8.2
1995	81.2	14.5
1996	79.8	12.9
1997	40.7	5.2
1998	106.7	19.4
1999	74.2	28.6
2000	37.3	9.0
2001	28.4	11.0
2002	54.4	12.9
2003	57.9	15.4
2004	104.6	41.6

Source: Munich Re.

Global production of crude steel reached 1.05 billion tons in 2004, an 8.8-percent increase over 2003.[1] (See Figure 1.) This was the first year in which steel output passed the billion-ton threshold. The surge occurred in an industry that is increasingly globally oriented and in which Asia, and especially China, plays a dominant role.

The steel industry has changed dramatically in the past decade. Japan was the leading producer in 1994, with China and the United States close behind.[2] Since then, China's output has nearly tripled, while output in Japan increased just 12 percent and U.S. production rose a mere 3 percent.[3] China, now the world leader, accounted for nearly half the increase in global production in 2004.[4] (See Figure 2.)

LINKS pp. 44, 46

Growth has been brisk in other Asian countries as well. While global steel production grew some 28 percent between 1994 and 2003, Asia's output increased by 65 percent.[5] Double-digit percentage increases in 2004 were registered by Thailand, Indonesia, and Kazakhstan as well as China.[6] Asia's share of total global production grew from 37 percent a decade ago to 48 percent in 2004, reflecting the economic dynamism of the region.[7]

Steel is increasingly a traded commodity. Exports of finished steel as a share of production have increased from roughly 25 percent in the mid-1980s to nearly 40 percent in 2001–03.[8] Key industrial sectors that drive the demand for steel are motor vehicles, construction, industrial machinery, fabricated metal products, and shipbuilding.[9] All are projected to grow steadily through 2010, with production of industrial machinery and fabricated metal products expected to grow fastest.[10]

Steel is one of the most extensively recycled materials, and demand for recycled steel is growing.[11] Global steel production in electric arc furnaces—steelmaking technology fed almost exclusively by scrap—increased by 37 percent between 1994 and 2003, outpacing total steel output.[12]

Cars are a common source of old steel scrap. Indeed, in 2003 the U.S. steel industry recycled more steel from cars than it used to produce new ones, giving autos a recycling rate of 103 percent.[13] Ninety percent of appliances and 60 percent of steel cans were recycled that year.[14] About half of the iron and steel scrap (the two materials are sometimes analyzed together) came from waste generated by consumers, while the rest came in equal parts from steel mills and factories that manufacture steel products.[15]

Steel consumption closely shadows economic growth in general, and China's hot economy is expected to make it the driver in global use in the near term. Consumption in China is expected to increase by more than 10 percent in 2005, and this one nation is projected to account for 61 percent of total growth in this year.[16] By comparison, growth in the rest of the world is expected to be just over 2 percent.[17]

China's appetite for steel may be affecting economies elsewhere. In November 2004, the Nissan Motor Company had to close three assembly plants in Japan for five days because of a lack of steel.[18] And a fire that shut down a mine in West Virginia that supplies coke, a kind of coal that fuels blast furnaces, led to production cutbacks at U.S. Steel because other supplies were unavailable in the tight coke market.[19]

The most widespread impact of Chinese steel consumption shows up in the price of steel and its inputs. Global steel prices jumped by 50–70 percent in the last half of 2003, to near-record levels.[20] The price of steel scrap, around $100 per ton in the 1999–2002 period, surged past $250 in 2004 as China increased its imports.[21] Chinese demand for coke has also risen sharply. A net exporter of coke until 2001, China now relies on imports for a growing share of its supply.[22]

China is working hard to secure its position in the global steel market by investing in steelmaking elsewhere. In 2004, for example, Minmetals Corporation led a consortium of Chinese state companies that sought to acquire Canada's largest mining company, Noranda, Inc.[23] This outward-oriented strategy, once shunned by Beijing in an effort to conserve foreign currency, is now encouraged by the government.[24]

Figure 1. World Steel Production, 1950–2004

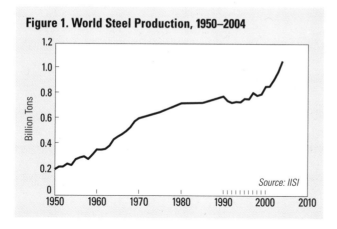

World Steel Production, 1950–2004

Year	Production
	(million tons)
1950	190
1955	270
1960	347
1965	450
1970	595
1975	644
1980	716
1985	719
1990	770
1991	734
1992	720
1993	728
1994	725
1995	752
1996	750
1997	799
1998	777
1999	788
2000	848
2001	850
2002	902
2003	965
2004	1,050

Source: IISI.

Figure 2. Top Five Steel-Producing Countries, 1994–2004

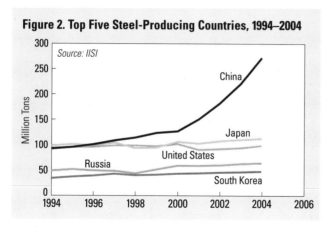

Transportation Trends

Battery installation in a gasoline-electric hybrid Toyota Prius

▶ Vehicle Production Sets New Record

▶ Bicycle Production Recovers

▶ Air Travel Slowly Recovering

Vehicle Production Sets New Record

Michael Renner

Global passenger car production grew 4.5 percent in 2004, to an estimated 44.1 million units.[1] Since 1950, annual car production has grown more than fivefold.[2] (See Figure 1.) Production of sport utility vehicles (SUVs) and other "light trucks" also reached a new record, 18 million, almost 6 percent more than in 2003.[3] There are now 551 million cars on the world's roads.[4] (See Figure 2.)

Car production and use remain heavily concentrated in North America, Western Europe, and Japan. The three regions together accounted for 70 percent of global passenger car production in 2003 and for more than two thirds of all cars on the roads in 2002.[5] Elsewhere, only Brazil, China, and South Korea are significant producers, and only Argentina, Australia, Brazil, India, Mexico, Poland, Russia, and South Korea have fleets of 5 million vehicles or more.[6]

LINKS pp. 40, 44, 94

Car density relative to population is by far the highest in the United States. Western Europe had a car density in 2002 comparable to the U.S. level of the 1970s.[7] And China's car density today is equivalent to U.S. levels in 1912.[8]

Automobiles are major contributors to global climate change. Carbon emissions from U.S. motor gasoline use—at 1,139 million tons in 2002—surpassed those of the entire Japanese economy.[9] Auto carbon emissions can be reduced significantly by boosting fuel efficiency. In the United States, fuel economy lags behind the levels reached in Japan and Europe.[10] Heavier cars, with more horsepower and faster acceleration, have prevented efficiency improvements during the last two decades.[11]

Despite the growing use of lightweight materials, average car weight has been on the rise since the mid-1980s and is now back to the level of the mid-1970s.[12] In 2003, a typical U.S. passenger car incorporated 824 kilograms of steel, 149 kilos of iron, 126 kilos of aluminum, and 116 kilos of plastics and composites.[13] Automobile manufacturing consumes huge quantities of materials, accounting for 33 percent of total U.S. aluminum use, 27 percent of iron, and 15 percent of steel.[14] (See Figure 3.)

In 1950, Americans drove some 588 billion kilometers (365 billion miles) in 40 million cars—almost 14,600 kilometers per car.[15] By 2003, the average distance traveled per year had grown to more than 19,000 kilometers (about 12,000 miles).[16] Multiplied by the far larger number of vehicles now on U.S. roads, the total distance traveled had thus grown more than sevenfold—to 4,281 billion kilometers.[17] That's the equivalent to 14,308 roundtrips from Earth to the sun.[18] Driving all these vehicles required 8.3 million barrels per day of fuel in 2002, up from 5.1 million barrels in 1970.[19] Passenger vehicle fuel consumption now surpasses total U.S. domestic oil production and is a major driver of rising imports.[20]

China is rapidly increasing its car dependency. Sales of cars and light commercial vehicles are expected to reach 5 million units in 2005 and 7.3 million by 2007.[21] The Chinese government introduced fuel economy standards for cars, SUVs, and minivans in late 2004.[22] These are more stringent than those prevalent in the United States but a bit less strict than the ones adopted semi-voluntarily by industry in Europe and Japan.[23]

Toyota and Honda are the pioneers in introducing hybrid electric cars (which complement the traditional internal combustion engine with an electrical motor, yielding lower fuel intake and less pollution). Worldwide, Toyota's cumulative hybrid sales surpassed 280,000 in late 2004.[24] The company expects that some 2 million hybrids will be sold by 2010.[25]

In the United States, Toyota and Honda have sold more than 120,000 hybrids since 1999.[26] Sales there are expected to reach some 200,000 units in 2005 alone.[27] With other carmakers getting on the bandwagon, analysts predict a continued doubling of hybrid sales, with perhaps 1 million hybrids on U.S. roads by 2007 or 2008.[28] Toyota expects hybrids to capture half the U.S. market for new cars by 2025, although this ambitious forecast is not universally shared.[29]

Hybrids and other alternative fuel vehicles still account for a marginal share of total car fleets. In the United States, an estimated 548,000 hybrids were on the roads in 2004, up from about 247,000 in 1995.[30]

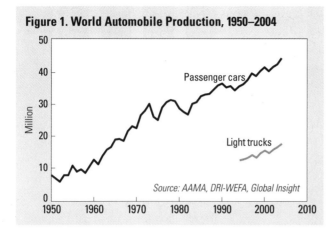

Figure 1. World Automobile Production, 1950–2004

Source: AAMA, DRI-WEFA, Global Insight

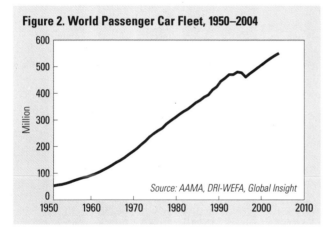

Figure 2. World Passenger Car Fleet, 1950–2004

Source: AAMA, DRI-WEFA, Global Insight

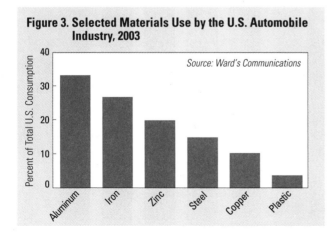

Figure 3. Selected Materials Use by the U.S. Automobile Industry, 2003

Source: Ward's Communications

World Automobile Production, 1950–2004

Year	Production
	(million)
1950	8.0
1955	11.0
1960	12.8
1965	19.0
1970	22.5
1971	26.5
1972	27.9
1973	30.0
1974	26.0
1975	25.0
1976	28.9
1977	30.5
1978	31.2
1979	30.8
1980	28.6
1981	27.5
1982	26.7
1983	30.0
1984	30.5
1985	32.4
1986	32.9
1987	33.1
1988	34.4
1989	35.7
1990	36.3
1991	35.1
1992	35.5
1993	34.2
1994	35.4
1995	36.1
1996	37.4
1997	39.4
1998	38.6
1999	40.1
2000	41.3
2001	40.1
2002	41.4
2003	42.2
2004 (prel)	44.1

Source: American Automobile Manufacturers Association, DRI-WEFA, and Global Insight.

Bicycle Production Recovers

Gary Gardner

Global bicycle production increased by more than 9 percent in 2002, the most recent year for which data are available, bringing production back up to 104 million—about the level of 2000.[1] (See Figure 1.) Output in most countries changed little or declined, but China produced 23 percent more bikes in 2002.[2]

Four of the top five producers are now Asian—China, India, Taiwan, and Japan.[3] (See Figure 2.) Viet Nam posted the world's fastest rate of growth, more than 250 percent, and produced more than 2 million bicycles.[4] The European Union, third in the world, is the only non-Asian producer of any size.[5]

LINKS pp. 44, 100

The industry has changed significantly over the past decade as production shifted steadily to China, which by 2002 accounted for 61 percent of the world total.[6] Once-large producers such as Japan, Taiwan, and the United States saw double-digit declines in output in 2002, part of a nearly decade-long trend in which cheaper and increasingly high-quality Chinese models grabbed market share worldwide.[7]

Although bicycles are only one segment of a society's transportation picture, their niche is underdeveloped nearly everywhere. The trend in many countries is toward greater automobile use, often at the expense of bikes. In several prospering Asian countries, for example, bicycles, rickshaws, and other forms of nonmotorized transport are being marginalized on city streets to make room for fast-growing car fleets.[8] And in the car-dominated United States, the share of trips to work by bike fell from 0.5 percent to an even more negligible 0.4 percent between 1980 and 2000.[9]

Bicycles are good for short-distance transport, for areas where nimble transportation is required, for users who cannot afford more expensive options, and for people seeking to combine commuting with exercise. Health care providers in Africa, for instance, have found that bicycles offer quick and inexpensive transportation. Bikes seem to be particularly effective in delivery of immunization programs, prenatal care, and ongoing therapies, such as the regime for treatment of tuberculosis.[10] Two projects in Senegal found that nurses using bikes were 58 percent quicker in their rounds than those who walked, and they saved 40¢ per trip over taking a taxi.[11]

Bicycles are also important complements to other forms of transportation. Bogotá, Colombia, may soon use bicycle taxis to provide feeder service to the stations of its metro-like bus system, bolstering the system's capacity to get citizens where they need to go cheaply and quickly.[12] And the city has installed safe, indoor bicycle parking facilities at some bus system stations to encourage riders to start their morning commutes on a bike.[13]

Boosting the bicycle's share of trips requires policies that shift incentives in its favor and that discourage car use. In the United States, where 95 percent of parking is free and where gas prices, vehicle taxes, and other driving-related costs are among the lowest in the industrial world, using a car is a rational choice and a key reason that biking remains marginalized.[14] The U.S. rate of car ownership is the highest in the world—and about 50 percent higher than in Western Europe.[15]

Safety is also a concern. In the United States, cyclists are 12 times more likely than people in cars to die en route.[16] On a per kilometer and per trip basis, U.S. cyclists are twice as likely to die on the road as German cyclists, and more than three times as likely as Dutch cyclists.[17] Cycling fatalities in these countries have fallen over the last quarter-century, but for very different reasons. U.S. cycling deaths have declined largely because of a drop in cycling, especially among children.[18] The Netherlands and Germany, on the other hand, have invested heavily in infrastructure that makes cycling safer.[19]

Six policies appear to have worked to promote cycling in Germany and the Netherlands: improved cycling infrastructure, "traffic calming" in residential neighborhoods, urban design that is people- rather than car-oriented, restrictions on motor vehicle use, traffic education, and traffic regulations and enforcement that are pro-pedestrian and pro-cycling.[20]

Figure 1. World Bicycle Production, 1950–2002

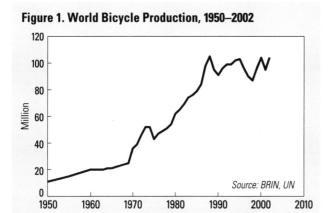

Source: BRIN, UN

Figure 2. Top Five Producers of Bicycles, 1990–2002

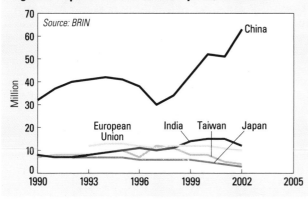

Source: BRIN

World Bicycle Production, 1950–2002

Year	Production
	(million)
1950	11
1955	15
1960	20
1965	21
1966	22
1967	23
1968	24
1969	25
1970	36
1971	39
1972	46
1973	52
1974	52
1975	43
1976	47
1977	49
1978	51
1979	54
1980	62
1981	65
1982	69
1983	74
1984	76
1985	79
1986	84
1987	98
1988	105
1989	95
1990	91
1991	96
1992	99
1993	99
1994	102
1995	103
1996	96
1997	90
1998	87
1999	96
2000	104
2001	95
2002	104

Source: Bicycle Retailer and Industry News and United Nations.

Air Travel Slowly Recovering

Zoë Chafe

World air travel rose less than 1 percent in 2003, the latest year with data available, according to the International Civil Aviation Organization (ICAO).[1] In 2003, passengers traveled 2.99 trillion passenger-kilometers, nearly recovering to levels posted before the unprecedented slowdown in air travel that followed the terrorist attacks of September 2001.[2] (See Figure 1.) Between 2000 and 2002, air travel fell by 73 billion passenger-kilometers, or 2.4 percent, from a high of 3.04 trillion passenger-kilometers.[3]

LINKS p. 40

In the 50 years since the first commercial jet was introduced, demand for air travel has increased by 9 percent a year on average, and the market is expected to continue growing over the next 20 years, albeit at only 3–5 percent per year.[4] Currently, 1.7 billion people (see Figure 2) and 35 million tons of freight are transported by aircraft each year.[5] North America generates just over one third of the global air traffic.[6]

The market for air travel is expanding rapidly in both the Asia/Pacific region and the Middle East.[7] Demand for domestic air transport in China is growing at the rate of 10 percent a year, compared with 2 percent a year in the United States.[8] In Africa, meanwhile, most aviation involves South Africa and is linked to either tourism or perishable food exports to Europe.[9]

Of 25,000 new planes slated for construction, approximately 17,000 will be for short-haul flights, which by 2023 are expected to account for 90 percent of all departures.[10] China's air fleet is due to skyrocket from 777 planes in 2003 to just over 2,800 planes in 2023.[11] Nearly two thirds of these are projected to be single-aisle planes, built for short-haul, usually domestic, routes.[12]

The world's airlines use some 205 million tons of aviation fuel (kerosene) each year, producing greenhouse gases such as carbon dioxide (CO_2), nitrogen oxides (NO_x), ozone, sulfur dioxide, and methane.[13] (Jet fuel is the second largest expense to airlines after labor and can amount to 20 percent of companies' operating expenses; one industry representative estimated that oil price increases in mid-2004 could add $1 billion a month to aviation costs.)[14] Aviation accounts for 2 percent of all human-caused CO_2 emissions but nearly all the NO_x emissions found 8–15 kilometers above Earth.[15]

Planes accounted for about 3.5 percent of the climate impacts due to human activities in 1992.[16] The Intergovernmental Panel on Climate Change estimates that by 2050, aviation could have 11 times as much impact on climate as it did in 1992.[17] The ICAO has been charged with coordinating the reduction of emissions from aircraft fuels, which are not covered by targets set in the Kyoto Protocol on climate change that went into effect in February 2005.[18]

Emissions from aviation can also produce contrails—clouds of water vapor, a greenhouse gas, that condense at high altitudes. After the September 2001 terrorist attacks, when nearly all aircraft were restricted from using U.S. airspace for several days, the difference between daytime and nighttime temperatures in the nation averaged 1–2 degrees Celsius above normal. This suggests that the absence of contrails lowered high cloud formation and allowed more sunlight to enter Earth's atmosphere, as well as providing less insulation against cooling at night.[19]

Planes use the most fuel—and produce the most harmful emissions—during takeoff. On short flights, as much as 25 percent of the total fuel consumed is used then.[20] Nearly three quarters of the new routes in Europe and North America are less than 2,000 kilometers long.[21] The most fuel-efficient length, however, is about 4,300 kilometers—roughly a flight from Europe to the U.S. East Coast.[22]

About 45 percent of all flights in the European Union cover less than 500 kilometers.[23] The Climate Action Network Europe estimates that a passenger traveling from Amsterdam to London would produce more than three times as much CO_2 traveling by plane than by train.[24] By improving rail systems, governments could provide a more sustainable alternative to the expected increase in short-haul air travel.[25]

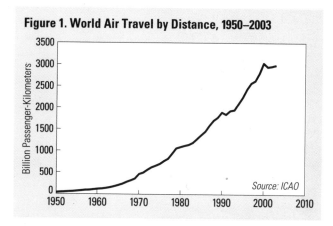

Figure 1. World Air Travel by Distance, 1950–2003

Source: ICAO

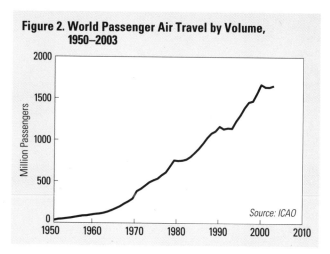

Figure 2. World Passenger Air Travel by Volume, 1950–2003

Source: ICAO

World Air Travel by Distance and Passenger Volume, 1950–2003

Year	Distance (billion passenger-kilometers)	Passengers (million)
1950	28	31
1955	61	68
1960	109	106
1965	198	177
1970	460	383
1971	494	411
1972	560	450
1973	618	489
1974	656	514
1975	697	534
1976	764	576
1977	818	610
1978	936	679
1979	1,060	754
1980	1,089	748
1981	1,119	752
1982	1,142	766
1983	1,190	798
1984	1,278	848
1985	1,367	899
1986	1,452	960
1987	1,589	1,028
1988	1,705	1,082
1989	1,774	1,109
1990	1,894	1,165
1991	1,845	1,135
1992	1,929	1,146
1993	1,949	1,142
1994	2,100	1,233
1995	2,248	1,304
1996	2,432	1,391
1997	2,573	1,457
1998	2,628	1,471
1999	2,798	1,562
2000	3,038	1,672
2001	2,950	1,640
2002	2,965	1,639
2003	2,992	1,657

Source: ICAO.

Health and Social Trends

Teenage boys listening outside the window of a sex education class, Mexico City

▶ Population Continues Its Steady Rise

▶ Number of Refugees Declines

▶ HIV/AIDS Crisis Worsening Worldwide

▶ Cigarette Production Drops

Population Continues Its Steady Rise
Danielle Nierenberg

The world's population grew to more than 6.3 billion in 2004, more than twice the number of people who were alive just 45 years ago.[1] (See Figure 1.) The global rate of population growth has actually decreased over the past three decades, from 2.1 percent a year in 1970 to 1.14 in 2004.[2] (See Figure 2.) But this does not mean the world's population is shrinking. In fact, in 2004 we added 73 million people to our numbers.[3] (See Figure 3.)

More than 95 percent of population growth occurs in developing countries, where fertility rates remain high.[4] Africa has the highest growth rate of any region, at 2.4 percent annually.[5] By 2050, the continent's population is projected to more than double, to 2.3 billion people.[6] Growth rates in Asia are lower, but they apply to a much larger base: the continent is now home to more than half the world's people.[7]

LINKS p. 108

But it is a mistake to think of population growth as a challenge facing only developing nations. In countries where population growth and high levels of consumption coincide, as they do in many industrial nations, the significance of added numbers of people balloons. Consider the United States and India. The U.S. population increases by roughly 3 million a year, while India's population increases by nearly 16 million.[8] The United States, however, has a significantly larger "ecological footprint"—it releases 15.7 million tons of carbon into the atmosphere each year, for instance, compared with India's 4.9 million tons.[9]

Other demographic trends also intersect with consumption in surprising ways. For instance, as a result of rising incomes, urbanization, and smaller families, the number of people around the world living under one roof declined between 1970 and 2000—from 5.1 to 4.4 in developing countries and from 3.2 to 2.5 in industrial countries—while the total number of households increased.[10] Each new house requires land and materials. When fewer people live in the same house, the savings gained from having more people share energy, appliances, and home furnishings are lost. A one-person household in the United States uses 17 percent more energy than a two-person household, for example.[11]

More and more of the world's people, particularly in the developing world, are moving to cities in search of jobs and other urban amenities. According to the United Nations, by 2007, more people will be living in cities than in rural areas.[12] Currently, five cities worldwide—Tokyo, Mexico City, New York, São Paolo, and Mumbai—have more than 15 million inhabitants.[13] Tokyo, with more than 35 million residents, is the largest city on Earth.[14]

Two other demographic challenges are occurring simultaneously. First, there are more young people than ever before. In 2000, more than 100 nations had populations where people aged 15 to 29 accounted for nearly half of all adults.[15] These youth bulges are all in the developing world, where fertility rates are the highest.[16] At the same time, however, many nations are concerned about their graying populations. In some countries—including Russia, Italy, and much of Eastern Europe—populations are actually declining. With a total fertility rate of barely more than one child per woman, Russia's population is now shrinking by 0.7 percent annually—roughly a million people a year.[17]

Lack of access to birth control and reproductive health services continue to prevent many people from planning their families. An estimated 350 million couples still lack access to a full range of family planning services, including contraceptives.[18] Almost 140 million women want to delay their next birth or avoid another pregnancy but are not using any family planning, and another 64 million women are using less effective methods.[19] Demand for family planning services is expected to increase 40 percent by 2025.[20]

Complications of pregnancy and childbirth are still one of the leading causes of illness and death among women in many parts of the developing world. According to the World Health Organization, 8 million women suffer life-threatening complications from pregnancy each year and more than 500,000 of them die.[21]

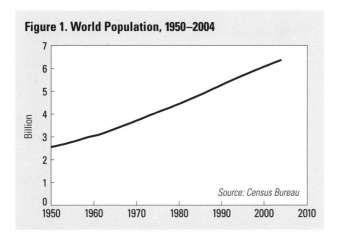

Figure 1. World Population, 1950–2004

Source: Census Bureau

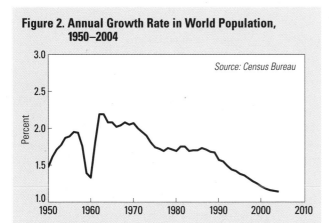

Figure 2. Annual Growth Rate in World Population, 1950–2004

Source: Census Bureau

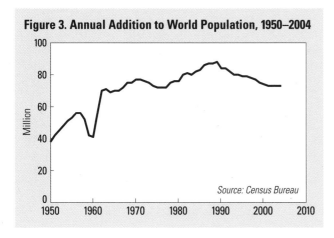

Figure 3. Annual Addition to World Population, 1950–2004

Source: Census Bureau

World Population, Total and Annual Addition, 1950–2004

Year	Total	Annual Addition
	(billion)	(million)
1950	2.56	38
1955	2.78	53
1960	3.04	41
1965	3.35	70
1970	3.71	77
1971	3.78	77
1972	3.86	76
1973	3.94	75
1974	4.01	73
1975	4.09	72
1976	4.16	72
1977	4.23	72
1978	4.30	75
1979	4.38	76
1980	4.45	76
1981	4.53	80
1982	4.61	81
1983	4.69	80
1984	4.77	82
1985	4.85	83
1986	4.93	86
1987	5.02	87
1988	5.11	87
1989	5.19	88
1990	5.28	84
1991	5.37	84
1992	5.45	82
1993	5.53	80
1994	5.61	80
1995	5.69	79
1996	5.77	79
1997	5.85	78
1998	5.93	77
1999	6.00	75
2000	6.08	74
2001	6.15	73
2002	6.23	73
2003	6.30	73
2004 (prel)	6.37	73

Source: U.S. Bureau of the Census.

Number of Refugees Declines

Michael Renner

The number of recognized international refugees declined to 9.7 million at the end of 2003, the most recent year with data.[1] (See Figure 1.) This is the lowest number since 1980, according to the U.N. High Commissioner for Refugees (UNHCR). Many refugees were able to return home after armed conflicts in Angola, Sierra Leone, and Liberia ended. In Afghanistan, the unseating of the oppressive Taliban regime allowed large-scale returns.

These numbers do not include Palestinians, however, as under a separate mandate they are covered by the United Nations Relief and Works Agency for Palestine Refugees in the Near East. The Palestinian refugee population has grown from 870,000 in 1953 to 4.2 million today.[2]

LINKS p. 74

During 2001–03, with UNHCR's help, some 4 million refugees voluntarily returned to their home countries, including Angola, Burundi, and Iraq.[3] The 645,000 Afghans were by far the largest group of returnees in 2003.[4] Another 760,000 Afghans returned during 2004, but 3 million more remained abroad—the second-largest group of refugees after the Palestinians.[5]

Other major sources of refugees are Sudan (606,000), Myanmar (586,000), Burundi (532,000), Democratic Republic of the Congo (453,000), and Somalia (402,000).[6] Iraqis, Vietnamese, Liberians, and Angolans—more than 300,000 in each group—are the next largest refugee groups.[7] The principal countries hosting refugees are Pakistan (home to 1.1 million refugees), Iran (985,000), Germany (960,000), Tanzania (650,000), and the United States (453,000).[8]

In addition to refugees, in 2003 UNHCR assisted close to 1 million asylum seekers, 1.1 million recent returnees who still needed assistance, nearly 1 million stateless persons (out of some 9 million stateless persons worldwide), and 4.4 million internally displaced persons.[9] Altogether, these "persons of concern" numbered 17.1 million at the end of 2003, down from 20.7 million in 2002.[10] (See Figure 2.)

The plight of the internally displaced is often far worse than that of recognized refugees. The Global IDP Project in Geneva notes that "in several cases, the protection of displaced people was undermined in the context of counterinsurgency campaigns intensified under the guise of the 'war on terror.'"[11] UNHCR can look after the internally displaced only with the consent of the national government in question. The 4.4 million people it helped in 2003 were a small portion of the global total.[12]

Figures by different sources show considerable variation. In 2003, the U.S. Committee for Refugees estimated there were close to 24 million internally displaced people, down slightly from 26.3 million in 2002.[13] (See Figure 3.) Half were in Africa, with some 5 million in Sudan alone, followed by 2 million in the Democratic Republic of the Congo and 1 million each in Angola and Uganda.[14] There were also 1 million displaced persons each in Myanmar, Iraq, and Turkey, and 2.7 million in Colombia.[15]

Some 3 million internally displaced people, most of them in Angola and Indonesia, were able to return home during 2003.[16]

The focus of UNHCR and similar agencies is on people uprooted by war and repression. But large numbers of people are driven from their homes, either temporarily or for good, by other factors—such as resource scarcity and unequal land distribution, climate change and other forms of environmental degradation, dams and other large-scale development projects, natural or industrial disasters, and destruction of the environment through warfare.

There is no generally agreed definition of environmental refugees, but Essam El-Hinnawi of the Natural Resources and Environmental Institute in Cairo estimates that perhaps 30 million people now fall into this category.[17] These numbers are likely to go up sharply in coming years. Desertification, for example, puts some 135 million people worldwide at risk of being driven from their lands.[18] And as climate change translates into more intense storms, flooding, heat waves, and droughts, more and more communities will likely be affected.

All in all, there may be at least 70 million displaced persons worldwide.[19] That translates into more than one out of every 100 persons on Earth.

Figure 1. International Refugees, 1951–2003

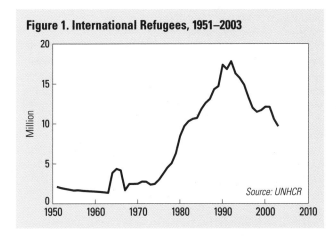

Figure 2. Refugees and Others of Concern to UNHCR, 1960–2003

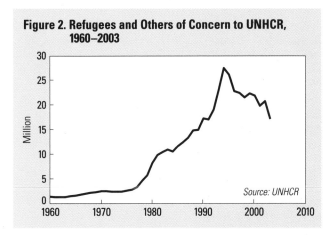

Figure 3. Internally Displaced Persons, 1986–2003

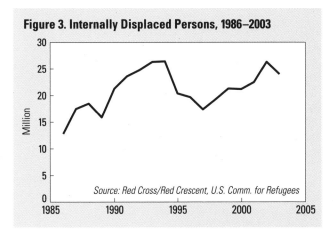

International Refugees, 1951–2003, and Internally Displaced Persons, 1986–2003

Year	Refugees	Displaced Persons
	(million)	
1951	2.1	
1955	1.6	
1960	1.5	
1965	4.4	
1970	2.5	
1971	2.8	
1972	2.7	
1973	2.4	
1974	2.5	
1975	3.0	
1976	3.8	
1977	4.5	
1978	5.1	
1979	6.3	
1980	8.4	
1981	9.7	
1982	10.3	
1983	10.6	
1984	10.7	
1985	11.9	
1986	12.6	12.8
1987	13.1	17.5
1988	14.3	18.5
1989	14.7	15.9
1990	17.4	21.3
1991	16.8	23.6
1992	17.8	24.8
1993	16.3	26.3
1994	15.7	26.4
1995	14.9	20.4
1996	13.4	19.7
1997	12.0	17.4
1998	11.5	19.3
1999	11.7	21.3
2000	12.1	21.2
2001	12.1	22.5
2002	10.6	26.3
2003	9.7	24.0

Source: UNHCR, Red Cross/Red Crescent, U.S. Committee for Refugees.

HIV/AIDS Crisis Worsening Worldwide   Lisa Mastny

In 2004, close to 5 million people were newly infected with human immunodeficiency virus, bringing to nearly 78 million the total number of HIV infections since the first AIDS cases were identified in 1981.[1] (See Figure 1.) Cumulative deaths from HIV-related illness grew by more than 3 million, to 34 million.[2] (See Figure 2.)

No disease in human experience debilitates and kills as AIDS does. Nearly 90 percent of fatalities occur among people of working age, making AIDS the leading cause of death worldwide for people ages 15 to 49.[3] The seven most seriously AIDS-affected countries—all in sub-Saharan Africa (see Figure 3)—now lose as much as 10–18 percent of their working-age adults every five years, mainly to this disease.[4] (Industrial countries, in comparison, typically lose about 1 percent of this age group to death in five years.)[5]

LINKS   p. 108

Largely because of this rising pandemic, death rates have actually reversed their decline in more than 30 countries worldwide.[6] At least 13 of the 53 countries now considered AIDS-affected have suffered measurable reversals in human development since 1990; in 7, life expectancy is less than 40 years.[7] Several of these countries could even see population declines soon as AIDS deaths overtake births.[8]

Where the epidemic is most advanced, the disease is widespread—affecting government, the armed forces, schools, factories, farms, and health care facilities.[9] Botswana and Zimbabwe, where more than a third of reproductive-age adults are HIV-positive, are among the hardest hit.[10] Botswana's largest diamond company, Debswana, suffered a tripling in AIDS deaths between 1996 and 1999.[11]

In perhaps 20 developing countries—nearly all of them in sub-Saharan Africa—more than 15 percent of the total military force is thought to be HIV-positive.[12] Some countries are experiencing military HIV infection rates that far exceed those among civilians: in Zimbabwe, an estimated three quarters of all soldiers now die of AIDS within a year of leaving the army.[13]

The International Labour Organization predicts that in the absence of treatment, as many as 74 million workers worldwide could die

from AIDS-related causes by 2015.[14] Employers in South Africa, home to the largest infected population, now face what economists term an AIDS "tax"—added expenditures for frequent sick leave, providing health care benefits and burial fees, and training new employees.[15] Between 1992 and 2002, the country's economy lost an estimated $7 billion annually due to declines in its labor force.[16]

Women and girls increasingly bear the HIV burden, as many become victims of their partners' high-risk behaviors. In 2004, the number of women living with AIDS worldwide reached 17.6 million, 45 percent of the world total.[17] Meanwhile, the number of children orphaned by the disease—the vast majority of them in Africa—increased from 11.5 million to 15 million between 2001 and 2003.[18]

A big wild card is how the disease will play out in China and India, where two fifths of the world lives and where HIV/AIDS surveillance efforts remain inadequate.[19] Although only about 1 percent of India's reproductive-age population is infected, some 5.1 million Indians now live with the disease, making it the second largest infected population in the world.[20] And because of China's mounting epidemic, the number of people living with HIV in East Asia jumped nearly 50 percent between 2002 and 2004, to 1.1 million.[21]

In Russia, rising intravenous drug use is contributing to the disease's rapid spread.[22] Without adequate prevention programs, according to the World Bank, as many as 650,000 Russians could be dying from HIV/AIDS annually by 2010—more people than have died of AIDS in the United States since 1981.[23]

Global funding for HIV/AIDS increased from some $2.1 billion to an estimated $6.1 billion between 2001 and 2004, and access to AIDS education and vital prevention and care services has improved greatly.[24] The number of people receiving antiretroviral therapy has jumped 56 percent since 2001, according to a survey of 73 developing countries.[25] Yet in many of the most affected countries, inadequate resources and a failure of political leadership continue to hamper progress.

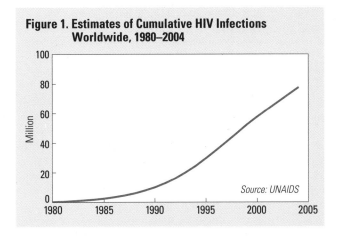

Figure 1. Estimates of Cumulative HIV Infections Worldwide, 1980–2004

Source: UNAIDS

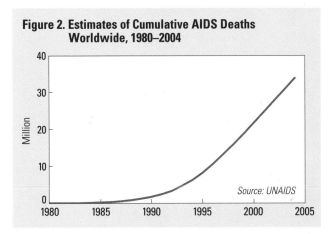

Figure 2. Estimates of Cumulative AIDS Deaths Worldwide, 1980–2004

Source: UNAIDS

Cumulative HIV Infections and AIDS Deaths Worldwide, 1980–2004

Year	HIV Infections	AIDS Deaths
	(million)	
1980	0.1	0.0
1981	0.3	0.0
1982	0.7	0.0
1983	1.2	0.0
1984	1.7	0.1
1985	2.4	0.2
1986	3.4	0.3
1987	4.5	0.5
1988	5.9	0.8
1989	7.8	1.2
1990	10.0	1.7
1991	12.8	2.4
1992	16.1	3.3
1993	20.1	4.7
1994	24.5	6.2
1995	29.8	8.2
1996	35.3	10.6
1997	40.9	13.2
1998	46.6	15.9
1999	52.6	18.8
2000	57.9	21.8
2001	62.9	24.8
2002	67.9	27.9
2003	72.9	30.9
2004 (prel)	77.8	34.0

Source: UNAIDS.

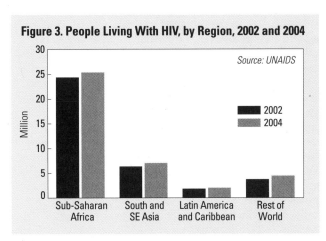

Figure 3. People Living With HIV, by Region, 2002 and 2004

Source: UNAIDS

- 2002
- 2004

Cigarette Production Drops

Erik Assadourian

After a small increase in 2003, global cigarette production declined 2.3 percent in 2004 to 5.5 trillion units.[1] (See Figure 1.) While total production has been about the same for the past decade, population growth during this period has reduced per capita output by 11 percent since 1994, to 868 cigarettes per person a year.[2] (See Figure 2.) Worldwide per capita production has not been this low since 1972.[3]

China, the United States, Russia, and Japan—the four largest producers—manufacture just over half of the world's supply.[4] In 2004, China produced 1.79 trillion cigarettes, 32 percent of the global total.[5] The United States produced 499 billion, 9 percent of the total.[6] This represents a substantial decline over years past. (In 1999, the United States produced 607 billion cigarettes, 22 percent more than in 2004.)[7] Unlike China, whose people smoked 99 percent of the cigarettes produced there, the United States exported 24 percent of its total production.[8]

Russia is currently the third largest producer, manufacturing 380 billion cigarettes in 2004—more than twice as many as in 1998.[9] Japan is the fourth largest, having produced 216 billion cigarettes.[10] Japan also imported a net 63 billion cigarettes, making this nation a leading smoker at 2,190 cigarettes per person—2.5 times the global average.[11] (See Figure 3.) But per person numbers mask the true smoking rates of most populations. In Japan, where 30 percent of people smoke, the average smoker actually goes through 7,228 cigarettes a year, about a pack a day.[12]

Today, 1.1 billion people smoke worldwide; 85 percent of these people live in low- or middle-income countries.[13] Future increases in smoking populations are expected to occur mainly in these regions, primarily because of higher rates of population growth and aggressive marketing campaigns by tobacco companies.[14]

Currently, smoking kills 4.8 million people a year prematurely—one in eight adults globally—mainly from cardiovascular diseases, chronic obstructive lung disease, and lung cancer.[15] Half of the victims live in industrial countries, and four fifths are men.[16] Experts project that smoking will become the world's leading cause of death by 2030, killing 10 million people a year—but by then 7 of every 10 fatalities will occur in low- or middle-income countries.[17]

Recent years have seen several success stories in local, national, and global efforts to curb smoking rates. In March 2004, Ireland became one of the first countries to ban smoking in all restaurants and bars.[18] In the first few months, tobacco sales fell 16 percent.[19] Several nations have already followed suit, with Norway implementing a ban three months later, and Italy doing so seven months after that; in Scotland, a ban will go into effect in 2006.[20] These laws have shown themselves to be effective at reducing both smoking and exposure to secondhand smoke.[21] In 1998, California was one of the first regions to adopt this form of smoking ban—a measure that has helped cut cigarette consumption in that state by 60 percent and the number of smokers by 27 percent.[22]

Poland, a country with one of the highest smoking rates in the world, has also cut cigarette consumption by 10 percent and the number of smokers by 29 percent.[23] In 1995 the Polish government passed legislation that banned sales to minors, severely restricted tobacco advertising, created prominent health warnings on cigarette packs, and prohibited smoking in enclosed workplaces.[24] This comprehensive response has helped to reduce annual smoking deaths by 10,000 and to decrease lung cancer rates by 30 percent among men aged 20 to 44.[25]

By far the most important tobacco control victory is the ratification of the Framework Convention on Tobacco Control (FCTC). This treaty entered into force on 27 February 2005, after having been ratified by over 40 countries.[26] The FCTC creates a strict international standard on tobacco control, obligating ratifying countries to increase tobacco taxes; ban advertising, sponsorship, and promotion; expand warning labels on tobacco products; increase protection from secondhand smoke; and implement measures to eliminate tobacco smuggling.[27]

Figure 1. World Cigarette Production, 1950–2004

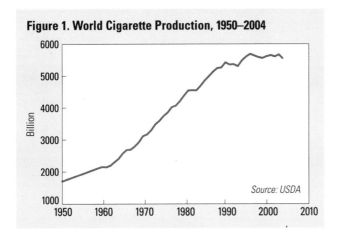

Source: USDA

Figure 2. World Cigarette Production Per Person, 1950–2004

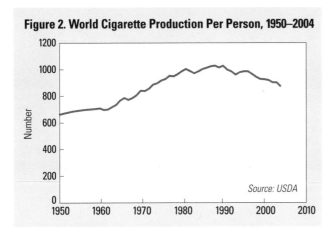

Source: USDA

Figure 3. Cigarette Consumption Per Person in the United States, China, and Japan, 1960–2004

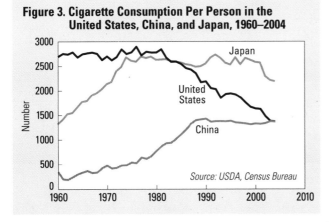

Source: USDA, Census Bureau

World Cigarette Production, 1950–2004

Year	Total (billion)	Per Person (number)
1950	1,686	660
1955	1,921	691
1960	2,150	707
1965	2,564	766
1970	3,112	839
1971	3,165	836
1972	3,295	853
1973	3,481	884
1974	3,590	894
1975	3,742	915
1976	3,852	926
1977	4,019	950
1978	4,072	946
1979	4,214	962
1980	4,388	985
1981	4,541	1,002
1982	4,550	987
1983	4,547	969
1984	4,689	983
1985	4,855	1,001
1986	4,987	1,011
1987	5,128	1,022
1988	5,240	1,027
1989	5,258	1,013
1990	5,419	1,026
1991	5,351	997
1992	5,363	984
1993	5,300	958
1994	5,478	976
1995	5,599	984
1996	5,680	984
1997	5,633	963
1998	5,581	941
1999	5,554	925
2000	5,609	923
2001	5,643	917
2002	5,602	900
2003	5,662	899
2004 (prel)	5,530	868

Source: USDA and U.S. Bureau of Census.

Conflict and Peace Trends

The U.S. Air Force's remotely piloted Predator, armed with a Hellfire missle

▶ Violent Conflicts Unchanged

▶ Military Expenditures Surge

▶ Peacekeeping Expenditures Soar

▶ Mixed Progress on Reducing Nuclear Arsenals

Violent Conflicts Unchanged

Michael Renner

According to AKUF, a conflict research group at the University of Hamburg, the number of wars worldwide stood at 26 in 2004.[1] In addition, there were 16 "armed conflicts" that were not of sufficient severity to meet AKUF's criteria for full-scale war.[2] Combining both categories, the total number of violent clashes stood at 42 in 2004, unchanged from the previous year.[3] (See Figure 1.)

Seven conflicts came to an end, but seven new ones erupted at the same time. Those terminated included conflicts in Macedonia, Central African Republic, Liberia, Democratic Republic (DR) of the Congo, Sri Lanka, Solomon Islands, and India-Pakistan.[4] New violence broke out in Haiti, the Kurdish region of Turkey, Georgia's South Ossetia province, Nigeria, Ethiopia, Yemen, and Thailand.[5]

LINKS pp. 76, 78, 80

A number of countries confronted multiple conflicts in their territories. There were six separate active conflicts in India, for instance, while Afghanistan, Colombia, DR Congo, Georgia, Indonesia, Nigeria, and the Philippines hosted two conflicts each.[6]

Information about armed conflicts is often incomplete or even contradictory. Because conflict researchers apply varying definitional and methodological tools to tally the number of armed conflicts, their findings often differ.[7] Researchers in Sweden and Norway, for instance, project a lower number than the AKUF group.[8] (See Figure 2.)

Researchers at the Heidelberg Institute for International Conflict Research in Germany are assessing conflict trends with a broader sweep. They found that out of 230 conflicts worldwide in 2004, 36 involved a high level of armed violence and 51 had occasional violence.[9] But another 143 disputes—62 percent—proceeded without the use of physical violence.[10] Most of the 230 active conflicts took place inside a given nation; only 66 were interstate conflicts.[11] The 25 conflicts counted by this group that escalated during 2004 were more than outweighed by 41 de-escalated cases.[12]

A variety of tools are being used to settle or otherwise address conflicts. In addition to an array of peacekeeping efforts, negotiations took place in 33 conflicts in 2004, resulting in four peace accords (in Burundi, Chad, DR Congo, and Sudan), six cease-fires (in Bhutan, Nepal, Sudan, and three separate times in Iraq), and 10 other agreements.[13] U.N. sanctions were maintained against six states—Afghanistan, Iraq, Liberia, Rwanda, Sierra Leone, and Somalia.[14] In 2003 and 2004, the International Court of Justice heard arguments in 26 disputes among states and handed down decisions in two cases each year.[15]

How many people perish in wars? Available information is often patchy or contradictory. Tallying combatant deaths is too narrow an approach, since both military personnel and civilians are killed in battle. The distinction between combatants and noncombatants may at times be blurred, and civilians (often accounting for the bulk of deaths in contemporary wars) may either be unintended "collateral damage" or be targeted intentionally. But in many conflicts a large share of the victims perish not because of battle wounds but due to the disease and starvation resulting from a collapsing economy and health system.[16]

Congo, Sudan, and Iraq are among the countries with the highest death tolls in recent years. In DR Congo, warfare since 1998 has caused more than 3.8 million people to perish, and more than 31,000 civilians continue to die each month.[17] (Battle deaths there ran to a comparatively small 145,000 in 1998–2001.)[18] Sudan's civil war since 1983 has killed an estimated 2 million people (but "only" 55,000 in battle).[19] Even as Sudan's north-south conflict wound down, some 70,000 people are believed to have died in 2004 in the country's Darfur region, a number that may well go up steeply.[20]

The British medical journal *The Lancet* published a study that estimated conservatively that the U.S.-British invasion of Iraq caused at least 98,000 deaths in the 17–18 months after the start of the war, many from air strikes.[21] The study concluded that the risk of death from violence after the invasion was 58 times higher than in the 14–16 months before the war.[22]

Figure 1. Wars and Armed Conflicts, 1950–2004

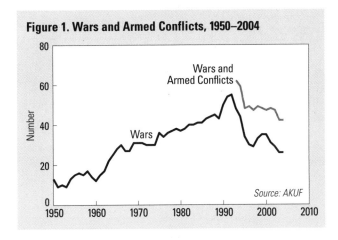

Figure 2. Armed Conflicts and Unclear Cases, 1950–2003

Wars and Armed Conflicts, 1950–2004

Year	Wars	Wars and Armed Conflicts
		(number)
1950	13	
1955	15	
1960	12	
1965	28	
1970	31	
1971	31	
1972	30	
1973	30	
1974	30	
1975	36	
1976	34	
1977	36	
1978	37	
1979	38	
1980	37	
1981	38	
1982	40	
1983	40	
1984	41	
1985	41	
1986	43	
1987	44	
1988	45	
1989	43	
1990	50	
1991	54	
1992	55	
1993	48	62
1994	44	59
1995	34	48
1996	30	49
1997	29	47
1998	33	49
1999	35	48
2000	35	47
2001	31	48
2002	29	47
2003	26	42
2004 (prel)	26	42

Source: AKUF and Institute for Political Science, University of Hamburg.

Military Expenditures Surge

Michael Renner

World military expenditures amounted to $932 billion in 2003, the most recent year for which data are available.[1] (See Figure 1.) Thus, every hour of every day the world spends more than $100 million on soldiers, weapons, and ammunition.[2] Following a steep decline from the cold war peak in the mid-1980s, about $200 billion has flowed back into military budgets since 1998.[3] In just two years, 2002 and 2003, spending rose by almost 20 percent.[4]

The United States is the planet's military colossus. It spends almost as much as all other countries on Earth combined, and the nation is without peer in terms of military technology or global reach.[5] The Bush administration's 2005 fiscal year request envisioned funding to rise from $421 billion in 2005 to $507 billion in 2009 (in inflation-adjusted dollars of 2003, from $401 billion to $440 billion).[6] (See Figure 2.) From 2010 onwards, current military plans would require budgets higher than those reached at the peak of cold war spending, and cost overruns for weapons procurement would likely push the numbers even higher.[7]

LINKS pp. 74, 80

The costs of the ongoing occupation of Iraq and of operations in Afghanistan are huge and mounting. The Bush administration has so far funded these through annual emergency supplementary appropriations rather than the regular budget process, and it intends to continue this practice.[8] Supplemental appropriations for Iraq and Afghanistan totaled $62.6 billion in FY 2003 and $65.6 billion in FY 2004.[9] In mid-2004, Congress agreed to appropriate another $25 billon, mostly for Iraq.[10] The administration is expected to request $81 billion more during calendar year 2005.[11]

The four largest spenders after the United States—Japan, the United Kingdom, France, and China—together accounted for 17 percent of global spending in 2003.[12] They were followed by Germany, Italy, Iran, Saudi Arabia, and South Korea, with a combined 12 percent share, and then by Russia, India, Israel, Turkey, and Brazil, with a combined 6 percent.[13] (Expressing budgets in dollars using market exchange rates tends to understate the purchasing power of developing and former Communist countries. Using purchasing power parity rates, China, India, and Russia are the largest spenders after the United States, but this measurement tends to overstate their military prowess.)[14]

High-income countries—home to only 16 percent of the world—account for $662 billion, or 75 percent, of global military expenditures.[15] Their military budgets are roughly 10 times larger than their combined development assistance.[16] In contrast, 58 low-income countries, with 41 percent of world population, account for just 4 percent of total military expenditures.[17] Even so, military spending represents a heavy burden for these impoverished nations, which also shoulder more than $500 billion in foreign debt.[18]

Spending large sums on the military and on the "war on terrorism" threatens to sideline international pledges agreed to in the Millennium Development Goals—pledges to counter poverty, health epidemics, and environmental degradation. Scarce financial resources and political capital are siphoned away from the root causes of insecurity.

Compared with military budgets, investments in health, education, and environmental protection are modest. Estimates suggest that programs to provide clean water and sewage systems would cost roughly $37 billion annually; to cut world hunger in half, $24 billion; to prevent soil erosion, another $24 billion; to provide reproductive health care for all women, $12 billion; to eradicate illiteracy, $5 billion; and to provide immunization for every child in the developing world, $3 billion.[19] Spending just $10 billion a year on a global HIV/AIDS program and $3 billion or so to control malaria in sub-Saharan Africa would save millions of lives.[20]

In 2003, donor countries gave $68 billion in official development assistance.[21] If all donors actually met their promises of providing 0.7 percent of their gross national income, annual development aid would increase by over $110 billion.[22]

Figure 1. World Military Expenditures, 1950–2003

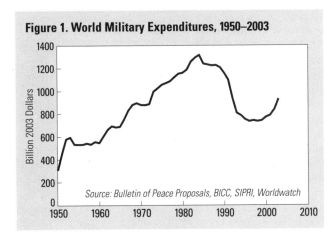

Source: Bulletin of Peace Proposals, BICC, SIPRI, Worldwatch

Figure 2. U.S. Military Expenditures: Budgeted Outlays (1950–2004) and Requests (2005–2009)

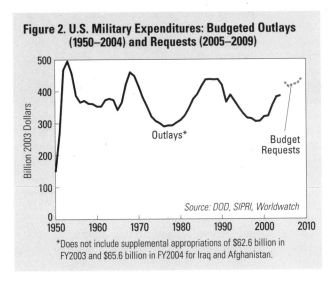

Outlays*

Budget Requests

Source: DOD, SIPRI, Worldwatch

*Does not include supplemental appropriations of $62.6 billion in FY2003 and $65.6 billion in FY2004 for Iraq and Afghanistan.

World (1950–2003) and U.S. (1950–2004) Military Budgeted Expenditures

Year	World	United States
	(billion 2003 dollars)	
1950	307	149
1955	532	387
1960	547	353
1965	688	342
1970	880	416
1971	878	380
1972	887	352
1973	995	321
1974	1,025	307
1975	1,056	300
1976	1,071	290
1977	1,087	293
1978	1,121	293
1979	1,152	302
1980	1,160	310
1981	1,187	324
1982	1,258	347
1983	1,292	375
1984	1,316	390
1985	1,240	415
1986	1,232	436
1987	1,224	437
1988	1,227	436
1989	1,208	437
1990	1,159	419
1991	1,099	366
1992	941	388
1993	808	368
1994	786	348
1995	751	330
1996	734	316
1997	741	314
1998	734	305
1999	741	306
2000	771	320
2001	787	322
2002	839	356
2003	932	382
2004	n.a.	385

Source: Tulberg, BICC, and SIPRI.

Peacekeeping Expenditures Soar

Michael Renner

Expenditures for United Nations peacekeeping operations from July 2004 to June 2005 are expected to soar to $3.8 billion—surpassing spending in the previous reporting period by $1 billion and nearing the 1994 peak.[1] (See Figure 1.) Peacekeeping staff also continued to grow rapidly. Some 64,720 soldiers, military observers, and police served in 16 peacekeeping missions as of December 2004.[2] (See Figure 2.) About 11,500 civilian staff also served in these missions.[3]

LINKS p. 74

In addition to full-fledged peacekeeping operations, the United Nations maintains 12 smaller "political and peace-building" missions, with a mostly civilian staff of 1,853.[4] By far the largest of these, with 938 workers, is the U.N. Assistance Mission in Afghanistan set up in March 2002.[5]

The 11 top contributors of personnel—among them Bangladesh, Pakistan, Nigeria, Ghana, and India—account for two thirds of all uniformed peacekeepers.[6] The permanent members of the Security Council—China, France, Russia, the United Kingdom, and the United States—provide less than 5 percent of personnel.[7] China's personnel commitment, however, has been rising rapidly, from fewer than 100 in 2001 to more than 1,000 at the end of 2004.[8]

New U.N. operations in Liberia, Côte d'Ivoire, Haiti, and Burundi initiated between September 2003 and June 2004 account for more than half of current expenditures and personnel.[9] The Liberia mission, with close to 16,900 military and civilian staff, is the largest operation, followed by the Democratic Republic of the Congo (with about 13,400 peacekeepers).[10] Missions in Haiti, Côte d'Ivoire, Burundi, Sierra Leone, and Kosovo each have between 5,000 and 8,000 peacekeepers.[11]

The numbers will continue to swell. The U.N. Security Council authorized another 5,900 peacekeepers to bolster its struggling Congo mission.[12] And U.N. Secretary-General Kofi Annan recommended that 10,700 peacekeepers be dispatched in support of a Sudanese peace agreement.[13]

U.N. peacekeeping continues to be plagued by financial problems. Between 2001 and 2003,

members reduced their outstanding arrears substantially. Yet 2004 saw a dramatic reversal, with members' debts surging from $1.1 billion to $2.57 billion in December 2004.[14] (See Figure 3.) Japan's debts ballooned to $759 million, and the country surpassed the United States as the single largest peacekeeping debtor.[15] Although Washington's debt rose from $482 million to $723 million in 2004, the U.S. share of total arrears actually fell from 45 to 28 percent.[16] France, China, Germany, and Italy each owe about $100 million, followed by South Korea with $70 million.[17]

In order to allow peacekeepers to be deployed more rapidly, the Association of Southeast Asian Nations is planning to build a regional peacekeeping force.[18] And the African Union is planning to set up a standby force by 2010.[19] The Group of Eight industrial countries have pledged training, equipment, and logistical support for this effort, envisioned to encompass some 50,000 peacekeepers.[20]

During 2004, some 30 non-U.N. peacekeeping missions were supported by regional organizations, such as the North Atlantic Treaty Organization (NATO), the Organization for Security and Co-operation in Europe, and the African Union, and by ad hoc coalitions of countries.[21] Altogether, non-U.N. peacekeeping operations involved about 60,000 soldiers in 2004.[22]

The largest of these are NATO-led multinational deployments in Bosnia, Kosovo, and Afghanistan.[23] The European Union (EU) is stepping up its involvement. In 2003, the EU sent small missions to Macedonia and eastern Congo, and in December 2004 it took over from NATO in Bosnia.[24] In addition, five EU countries decided to create a joint 3,000-strong European Gendarmerie Force for post-conflict tasks.[25]

Traditional military deployments abroad still dwarf peacekeeping efforts. Some 530,000 soldiers (70 percent of them from the United States) in such operations overshadow the 125,000 U.N. and non-U.N. peacekeepers worldwide.[26] The Iraq occupation alone involved some 170,000 soldiers from the United States and its allies during 2004.[27]

Figure 1. U.N. Peacekeeping Expenditures, 1950–2004

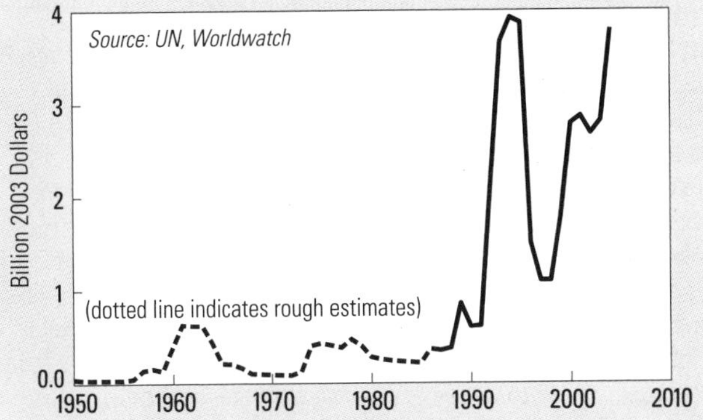

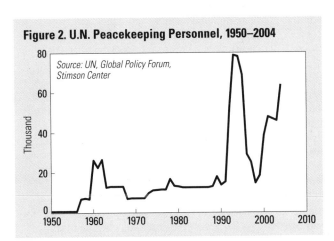

Source: UN, Worldwatch

(dotted line indicates rough estimates)

Figure 2. U.N. Peacekeeping Personnel, 1950–2004

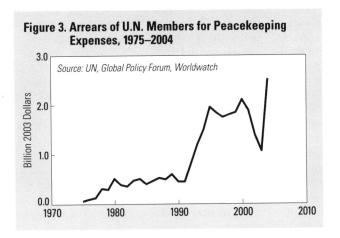

Source: UN, Global Policy Forum, Stimson Center

U.N. Peacekeeping Expenditures, 1986–2004

Year	Expenditure
	(billion 2003 dollars)
1986	0.360
1987	0.348
1988	0.373
1989	0.856
1990	0.603
1991	0.615
1992	2.168
1993	3.669
1994	3.924
1995	3.871
1996*	1.502
1997*	1.100
1998*	1.099
1999*	1.784
2000*	2.788
2001*	2.867
2002*	2.682
2003*	2.820
2004* (prel)	3.801

* July to June of following year.

Source: U.N. Department of Public Information and Worldwatch Institute database.

Figure 3. Arrears of U.N. Members for Peacekeeping Expenses, 1975–2004

Source: UN, Global Policy Forum, Worldwatch

Mixed Progress on Reducing Nuclear Arsenals *Michael Renner*

The world's five full-fledged nuclear powers—the United States, Russia, the United Kingdom, France, and China—together held about 19,000 nuclear warheads in 2004.[1] (See Figure 1.) In addition, some 9,000–10,000 Russian warheads are in storage or await dismantlement, although reliable annual numbers on these are not available.[2] Including "spares" (weapons in storage) as well as plutonium cores in reserve, the five powers still possess some 36,500 warheads.[3]

The United States and Russia control about 95 percent of global nuclear arsenals.[4] France, China, and the United Kingdom have roughly 1,000 warheads combined.[5] (See Figure 2.) Israel, India, and Pakistan have joined the nuclear "club" but have not signed the Nuclear Non-Proliferation Treaty (NPT). Israel is widely believed to have 100–200 warheads.[6] India is thought to possess 30–40 warheads and Pakistan, 30–50.[7]

LINKS pp. 74, 76

The United States and Russia signed a Strategic Offensive Reduction Treaty in May 2002 that commits them to reduce their deployed warheads to 1,700–2,200 each by December 2012.[8] But unlike earlier nuclear treaties, the agreement does not provide for verification or require that surplus warheads actually be dismantled.[9]

Since 1945, some 128,000 warheads have been built—about 70,000 by the United States; 55,000 by the Soviet Union/Russia; 1,200 by the United Kingdom; at least 1,260 by France; and some 600 by China.[10]

It takes 8 kilograms of plutonium or 25 kilograms of highly enriched uranium (HEU) to produce a nuclear weapon.[11] The five nuclear powers have halted their production of fissile material, but Israel, India, and Pakistan have not.[12] Stocks of HEU worldwide come to 1,900 tons.[13] (See Figure 3.) Global stocks of weapons-grade plutonium amounted to at least 262 tons at the end of 2003.[14] Far more plutonium is found in the fuel rods of civilian nuclear power plants: some 1,595 tons, growing by 70–75 tons a year.[15] (The greatest risk in terms of possible weapons use lies with the 235 tons of civilian plutonium that has not been irradiated.)[16]

A treaty outlawing production of weapons-grade fissile materials, though proposed for many years, remains controversial. And in July 2004, the United States announced that it would oppose the inclusion of any inspection and verification provisions in such a treaty.[17]

The NPT obligates the five nuclear powers to work toward the elimination of their arsenals. Yet they are not only intent on retaining these weapons; they are also modernizing them and, except for the United Kingdom, developing new systems.[18] The U.S. 2001 Nuclear Posture Review asserted that nuclear weapons "provide credible military options to deter a wide range of threats" and help "achieve strategic and political objectives."[19] The United States is seeking to develop more usable warhead concepts and designs, including low-yield nuclear weapons and earth-penetrating warheads capable of destroying adversaries' underground facilities.[20] Having provided funding for this in 2003, however, Congress rejected administration budget requests in November 2004.[21]

Yet the existing powers are determined to uphold one aspect of the NPT—barring other countries from acquiring nuclear weapons. They have adopted export controls, sanctions, and measures like the Proliferation Security Initiative, under which the United States and several key allies are intercepting planes and ships suspected of carrying nuclear, chemical, or biological weapons or missile components.[22]

Following several months of secret diplomacy, Libya announced in December 2003 that it would dismantle its nuclear program under international inspection.[23] But concerns about other countries' plans have risen in the last few years, particularly after it was revealed that scientist A.Q. Khan of Pakistan ran an illicit network aiding the efforts of Libya, North Korea, Iran, and possibly others to acquire nuclear weapons technologies.[24]

North Korea has apparently maintained a secret program to produce weapons-grade uranium.[25] It is also believed to have produced weapons-grade plutonium, but the government has adopted a policy of deliberate ambiguity about its nuclear status.[26] Iran, too, may be in pursuit of a nuclear-weapons capability.[27]

Figure 1. Global Nuclear Warheads, 1945–2004

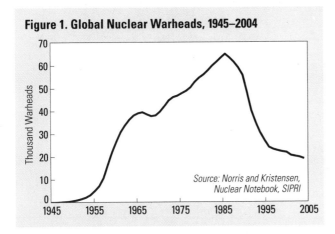

Source: Norris and Kristensen,
Nuclear Notebook, SIPRI

Figure 2. Nuclear Warheads in China, France, and the United Kingdom

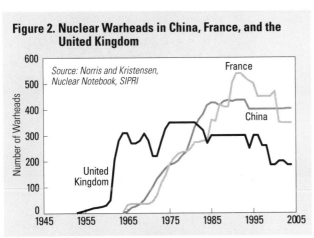

Source: Norris and Kristensen,
Nuclear Notebook, SIPRI

France

China

United Kingdom

Figure 3. Global Stocks of Plutonium and Highly Enriched Uranium, 2003

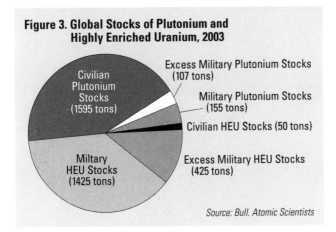

Civilian Plutonium Stocks (1595 tons)

Excess Military Plutonium Stocks (107 tons)

Military Plutonium Stocks (155 tons)

Civilian HEU Stocks (50 tons)

Miltary HEU Stocks (1425 tons)

Excess Military HEU Stocks (425 tons)

Source: Bull. Atomic Scientists

Global Nuclear Warheads, 1945–2004

Year	Warheads
1945	6
1950	374
1955	3,267
1960	22,069
1965	38,118
1970	38,153
1971	39,822
1972	42,194
1973	44,791
1974	46,195
1975	46,830
1976	47,913
1977	48,921
1978	50,441
1979	52,862
1980	54,706
1981	56,034
1982	57,858
1983	59,937
1984	61,624
1985	63,417
1986	65,057
1987	63,484
1988	61,550
1989	59,124
1990	55,863
1991	48,176
1992	40,161
1993	34,897
1994	30,571
1995	27,131
1996	24,121
1997	23,203
1998	22,637
1999	22,184
2000	21,871
2001	20,567
2002	20,148
2003	19,767
2004 (prel)	19,095

Source: Norris and Kristensen, "Nuclear Notebook," and SIPRI.

Part Two

SPECIAL FEATURES

Environment Features

Digital Vision

Logging trucks in Sabah, Borneo

▶ Mammals in Decline

▶ Global Ice Melting Accelerating

▶ Wetlands Drying Up

▶ Forest Loss Continues

▶ Air Pollution Still a Problem

Mammals in Decline

Howard Youth

Nearly one in four mammal species is in serious decline, mainly due to human activities.[1] (See Table 1.) Within the next five decades, such well-known animals as chimpanzees—primates with which humans share 98 percent of their DNA—may be extinct in the wild.[2] To date, 73 mammal species are known to be extinct; 4 more are extinct in the wild— found now only in zoos and other breeding facilities.[3] Many of the remaining species survive in vestiges of their former ranges. Africa's once-widespread black and white rhinoceroses provide one example. Today, about 80 percent live in South Africa, while many nations where rhino traditionally roamed no longer have any.[4] Between 1970 and 1992, black rhino numbers plummeted 96 percent due to hunting.[5]

LINKS p. 64

Mammals quickly become isolated when new roads, settlements, farms, or logging operations carve up their habitats. Many species, such as tigers and China's giant panda—now found only in 24 patches of mountainous forest—live in populations peppered across heavily farmed, increasingly populated areas, few of which are large enough to sustain these animals well into the future.[6]

Declines, particularly in large animals, can also be seen at the family level: 48 percent of wild cat species have critically endangered, endangered, or vulnerable status, as do 50 percent of bears, four of five rhinoceros species, both elephant species, and all great ape species (chimpanzees, gorillas, and orang-utans).[7] Many more species fall into lower categories of concern, including near-threatened and "data-deficient," a category that reflects how poorly studied many mammals remain.

Several factors contribute to mammal population losses, virtually all of them driven by human actions. The most widespread problems are habitat loss and habitat fragmentation, which are often compounded by uncontrolled hunting. This combination quickly kills or drives off the largest mammals.[8] Hunting provides the most immediate threat to large animals such as rhinoceroses, elephants, tapirs, jaguars, and many primates.[9]

Although cultures around the world celebrate various large mammals—bears, for example, are often revered for their human-like appearance, power, and intelligence—such respect rarely translates into action that lets these animals live alongside humans.[10] In fact, people also value bears and other animals for their parts: in China and other Asian countries, bear gall bladders, rhinoceros horn, and tiger bones and penises are prized in a surging traditional medicine market that fuels declines in wildlife populations already stressed by overhunting and habitat loss.[11]

Hunters also target a wide variety of mammals for food, not just for subsistence purposes but also to feed growing "bushmeat" markets in Africa and Asia. Meanwhile, illegal trade in animals for their skins and for the pet trade is rife in many countries such as Indonesia.[12] Despite international agreements to protect species, few countries have resources to pursue poachers or illegal market trading in wildlife. In recent years, resources for wildlife protection have shrunk in Central and West Africa and in Asia, where many elephants are being illegally killed for their ivory and meat.[13]

Their cultural, medicinal, and monetary value aside, mammals play vital roles within their ecosystems. Bats pollinate flowers and control insect pests, predators keep deer and rodent populations in check, and a wide variety of plant-eating mammals—from rodents to tapirs—help disperse seeds of native plants.[14] Loss of such services can cause grave imbalances within ecosystems.[15] After the decimation of cougar and wolf populations in eastern North America, for example, white-tailed deer populations surged; especially when they are hemmed in by habitat fragmentation, these animals chew down forest undergrowth, destroying native plants and habitat for nesting songbirds and other wildlife.[16]

Just keeping clear of human traffic often proves impossible for endangered mammals. Hunted by whalers for centuries, the North Atlantic right whale is now killed by ship collisions and fishing gear entanglements, fatalities that further threaten the remaining population of about 300 animals.[17] On land, an ever-

Table 1. Conservation Status of the World's Mammals, 2004

Species Status	Number
Extinct	73
Extinct in the Wild	4
Critically Endangered	162
Endangered	352
Vulnerable	587
Lower Risk/Conservation Dependent	64
Near-Threatened	587
Data-Deficient	380
Least Concern	2,644
Total	4,853

Source: IUCN.

expanding web of roadways proves fatal for endangered and declining mammals and provides physical barriers that isolate other mammals that will not cross clearings.[18]

The world's changing climate is emerging as a new challenge for mammal populations. The polar bear may be one of the first victims. In recent years, sea ice has melted earlier in the year in areas such as Hudson Bay, making it more difficult for bears to hunt seals, their primary prey.[19]

In other areas, too little water exacerbates the dangers mammals face. Once a fixture on East Africa's arid and semiarid plains, the Grevy's zebra is now endangered, its breeding success cut short after critical water sources dried up due to irrigation schemes or became crowded with cattle herds, which forced the zebras to drink at night, when they are more vulnerable to predators.[20]

Introduced species also menace mammal populations, either by spreading pathogens, through predation, or through direct competition. Village dogs serve as vectors of rabies and other diseases that further threaten endangered Ethiopian wolf and African wild dog populations.[21] Across Australia, introduced red foxes and feral cats feed on threatened native mammals, while in Europe, introduced eastern gray squirrels and American mink push native squirrels and mink out of their habitat.[22]

There is no blanket fix to stem the loss of mammal populations. Each species, each habitat, faces its own challenges and requires careful study before aggressive action.[23] Often, effective conservation can be coupled with human activities. The problem is that in many areas, and for most mammal species, too little is known of the animals' biology and habitat needs to take effective steps.[24] In fact, new mammal species continue to be discovered, particularly in Amazonia and other tropical forest areas.[25]

In areas such as Tanzanian game parks, eastern North American forests, western European forests, and Asian tiger reserves, wildlife corridors are needed to link core areas so that populations can mingle using vegetated pathways. In addition, in order for large mammals to survive, local communities must be involved in conservation efforts that protect large blocks of habitat, while embracing and profiting from land uses that serve both human and wildlife needs.[26]

Defining just how large habitat reserves must be is one of the great challenges in balancing mixed land use and conservation.[27] Even with proper management, however, parks cannot adequately protect many species. Most threatened mammals live outside of declared reserves, often in patchworks of remnant habitat and agricultural land.[28] For example, just 12–14 percent of the tiger's remaining range falls under protected status, while for the lion the figure is 9–12 percent and for the jaguar, 3–6 percent.[29] Recent studies indicate that many animals can move among habitat patches via agricultural landscapes, including coffee plantations and vineyards, if the mosaic of wild and settled habitats retains enough variety.[30]

In the years ahead, local involvement along with careful study, planning, and implementation will determine the future of the world's mammal diversity. Without these elements working in tandem, the prospects for many species will worsen.

Global Ice Melting Accelerating

Lisa Mastny

From the polar regions to high mountain glaciers, Earth's ice cover is melting at a rapid rate.[1] (See Table 1.) Global ice melt accelerated during the 1990s, which was also the warmest decade on record.[2] Scientists suspect that the enhanced melting is related to the unprecedented release of greenhouse gases by humans during the past century.[3]

p. 40

Some of the greatest melting is at Earth's poles. Over the past 50 years, temperatures in the Arctic have risen at nearly twice the global average, with parts of Alaska and Siberia warming even faster.[4] The impact on the region's sea ice has been dramatic: in 2002, coverage was 15 percent below average—representing a loss in area nearly twice the size of Iraq.[5] Scientists project that summer sea ice in the Arctic Ocean could disappear by the end of the century, opening the region to new shipping and oil opportunities as well as to greater ecological risks.[6]

Arctic warming is also affecting Greenland, the northern hemisphere's largest land-based ice area and the site of 10 percent of Earth's frozen water.[7] Studies show that the flow of glacial ice on the island has accelerated in recent years, and the melt zone has expanded further inland.[8]

The massive Antarctic ice cover, which averages 2.3 kilometers in thickness and represents some 91 percent of Earth's ice, is melting as well.[9] So far, most of the loss has occurred along the edges of the Antarctic Peninsula, where temperatures have risen by roughly 2.5 degrees Celsius over the past half-century.[10] In the last 30 years, at least 13,500 square kilometers of peninsular ice shelves have disintegrated, including the Prince Gustav and Larsen A and B shelves during just the past decade.[11]

Antarctica's two massive ice sheets—home to 70 percent of Earth's fresh water—may be starting to weaken.[12] Inland glaciers once held up by the now-shattered 3,250-square-kilometer Larsen B ice shelf, are flowing into the Weddell Sea up to eight times faster than before—suggesting that continued loss of the shelves could speed the ice cap's overall demise.[13]

Outside the poles, most ice melt has occurred in the world's mountain and subpolar glaciers, which are extremely sensitive to temperature shifts. According to the World Glacier Monitoring Service, glaciers experienced "spectacular loss in length, area and volume" in the twentieth century, and most are now shrinking faster than they are growing.[14] By one estimate, the world's glaciers lose at least 90 cubic kilometers of ice annually—as much water as all U.S. homes, factories, and farms use every four months.[15]

Melting glaciers and ice sheets have contributed a growing share of the estimated global sea level rise of 10–25 centimeters over the past century.[16] (Loss of sea ice and of floating ice shelves has no effect on ocean levels because these already displace water.) Ice melt could now account for more than half of all sea level rise, with rapidly retreating Alaskan glaciers contributing the largest share.[17]

The disappearance of Earth's ice cover would significantly alter the global climate, though the net effect remains unknown. Already, ice melt is accelerating warming at the poles by exposing darker land and water surfaces that absorb heat.[18] Increased Arctic thawing could also release large amounts of carbon dioxide and other climate-altering greenhouse gases from the frozen tundra.[19] But excessive Arctic melt may have a cooling effect in parts of Europe and the eastern United States if the influx of fresh water into the North Atlantic disrupts ocean circulation patterns that enable the warm Gulf Stream to flow north.[20]

As mountain glaciers disappear, large regions that rely on glacial runoff for water could face severe shortages. Glaciers in Bolivia's Cordillera Real, which supply 1.5 million people in La Paz and El Alto with water, are now losing mass at much faster rates than in previous decades.[21] And if Nepal's Himalayan glaciers dry up, rivers that feed the Ganges River in downstream India could see flow reductions of up to 90 percent.[22]

Rapid melting creates other hazards as well. In Alaska, the Inuit village of Shishmaref has lost more than 90 meters of coastline to related flooding and erosion in the past 30 years—half of it since 1997—and moving residents to more stable ground will cost an estimated $100 million.[23] In September 2002, the collapse of the Maili glacier and a 20-million cubic meter gla-

Table 1. Selected Examples of Ice Melt Around the World

Name	Location	Measured Loss
Arctic Sea Ice	Arctic Ocean	Has shrunk by some 8 percent (an area larger than Denmark, Sweden, and Norway combined) over the past 30 years, and the melting is accelerating. Has thinned by 15–20 percent overall since the late 1960s, with losses in some areas near 40 percent.
Greenland Ice Sheet	Greenland	Surface melt area increased 16 percent between 1979 and 2002—to a record 685,000 square kilometers. Margins are now melting by as much as 10 meters per year, 10 times faster than in 2001. Speed of flow of the largest outlet glacier has doubled since 1997, to 12.6 kilometers per year.
Glaciers	Alaska, United States	Now thinning by 1.8 meters a year on average, more than twice the annual rate observed from the 1950s to the mid-1990s. Total ice loss is estimated at 96 cubic kilometers each year.
Glacier National Park	Montana, United States	Since 1850, the number of glaciers has dropped from 150 to less than 40. The park's remaining glaciers could disappear completely in 30 years.
Central and Southern Andes	Peru	Have lost 20 percent of their 2,600 kilometers of glaciers in the past 30 years. All 18 glacier-capped mountains are now melting. At current rates, glaciers below 5,500 meters could disappear by 2015.
Patagonian Ice Fields	Argentina and Chile	Northern and southern fields have been in retreat for roughly a century. Average rate of thinning was twice as fast between 1995 and 2000 as between 1975 and 2000. Now lose 42 cubic kilometers of ice a year.
Amundsen Sea area	West Antarctica	Glaciers discharge some 250 cubic kilometers of ice each year, nearly 60 percent more than they accumulate from inland snowfall. Thinning rates in 2002–03 were much higher than during the 1990s.
Tibetan glaciers	China	Retreated 7.5 percent between 1850 and 1960 and a further 7 percent in the following 40 years. In the 1990s alone, the glaciers shrank by more than 4 percent.
Himalayas	Nepal	Average snow and ice cover in the east has decreased by 30 percent in the last 30 years. Within the next 35 years, the total glacial area is expected to shrink by one fifth, to 100,000 square kilometers.
Mount Kilimanjaro	Tanzania	Lost 82 percent of its ice between 1912 and 2000, shrinking from 12 square kilometers to 2.6 square kilometers. Ice could disappear completely by 2015.
The Alps	Europe	Glaciers lost roughly one third of their area and one half their mass between 1850 and 1980. Since 1980, a further 20–30 percent of the remaining ice has melted. The hot summer of 2003 alone accounted for a large share of this loss: Swiss glaciers retreated 3 meters in 2003, compared with 70 centimeters a year in the 1990s. Three quarters of Swiss Alpine glaciers are projected to disappear by 2050.

Source: See endnote 1.

cial lake in the Caucasus sent 3 million tons of ice and rock down slopes, killing at least 17 people and leaving a 33-kilometer swath of mud, ice, and debris.[24] The United Nations reports that as many as 40 glacial lakes in the Himalayas could burst in the next 5–10 years.[25]

Wildlife, too, is suffering as a result of global ice melt.[26] Changes in ice cover in northern Canada have led to hunger and weight loss among polar bears, and continued thawing could affect populations of migratory birds that breed in the high Arctic.[27] In Antarctica, loss of the sea ice, together with rising air temperatures and increased precipitation, has altered the habitats as well as feeding and breeding patterns of penguins and seals.[28]

Wetlands Drying Up

Howard Youth

Wetlands cover up to 6 percent of Earth's surface but provide a disproportionate amount of natural goods and services.[1] They vary in character, depending on the natural processes that shape them, and range from marshes and mudflats to mangrove forests, ponds, swamps, wet meadows, and bogs. All are characterized by the major role water plays in their ecology.

The world's wetlands harbor staggering biodiversity, protect vital water supplies and fisheries, and provide medicinal, agricultural, and timber products. In addition, they buffer coastal or riverside areas from storms and floods, control erosion, facilitate groundwater recharge and discharge, help maintain water quality, and retain nutrients and sediments. Many are also valued for their recreation and tourism opportunities.[2]

LINKS p. 26

Despite these assets, an estimated half of the world's wetlands have been lost since 1900, and their destruction continues apace.[3] The main causes of this loss have been drainage and conversion of wetlands to agricultural or urban land, compounded by pollution.[4]

The Mesopotamian marshlands of Iraq and western Iran—the largest remaining wetland ecosystem in the Middle East and western Eurasia—provide a particularly striking glimpse at wetland destruction.[5] By 2000, more than 90 percent of this unique ecosystem had dried up (see Table 1) after construction of dams upstream followed by concerted efforts by the Iraqi government between 1991 and 1997 to drain the wetlands.[6] Many wildlife populations were wiped out, and most of the area's indigenous Marsh Arabs were forced to abandon their land.[7]

As in many other wetlands around the world, the drawdown of fresh water also brought increased salinity to the Mesopotamian marshes, changing plant composition, ruining nearby cropland, and compromising the wetlands' ability to regenerate in many areas.[8] Blowing sediment and salt now contribute to growing health problems, while pollution caused by bombs, oil spills, and the destruction of local industries and sanitation facilities further threaten communities and remaining wetlands.[9]

This environmental catastrophe rivals the drastic reduction in the Aral Sea's once-extensive wetlands in Central Asia. After the Amu Dar'ya and Syr Dar'ya were diverted in the 1960s to feed local irrigation systems, the lake—once the world's fourth largest—shrank by half, cutting its volume by two thirds and increasing its salinity.[10] Smaller wetland losses have also resulted in large environmental troubles. In Armenia, for example, water diversion efforts drastically lowered water levels of Lake Sevan and dried up Lake Gilli by 2000; as a result, 27 bird species stopped nesting in the area.[11]

Many wetlands occur in border areas or are fed by water sources in different countries, making their conservation a truly international challenge. Damming and extensive irrigation schemes, for instance, exacerbated the effects of drought on the Sistan wetlands at the border with Iran and Afghanistan. Since 1998, 99 percent of this area has dried up, including the Iranian portion declared a wetland of international importance under the Ramsar Convention.[12] In Iraq, any future marsh restoration likely will prove impossible without cooperation from Turkey and Syria, countries whose dams now affect water flow in the Tigris and Euphrates watersheds.[13] Meanwhile, Chinese dams operating on the Mekong River may already be stressing fisheries and wetlands as far downstream as Cambodia.[14]

Even wetland areas protected as reserves or parks cannot escape the effects of activities outside their borders. In the 1960s, Spain's Las Tablas de Daimiel was a semiarid wetland spanning 6,000 hectares.[15] Today, just 1,928 hectares remain, protected as a national park.[16] Due to water diversions and extensive irrigation of surrounding farmlands, the wetland must now be artificially inundated through an aqueduct, and even then only part of it receives water. Increased pollution and salinity have accompanied decreased water inputs, dramatically changing the ecosystem's plant and animal life over the last 35 years.[17]

An advancing menagerie of introduced species compounds the problems wetlands face

Table 1. Loss in Selected Wetlands

Wetland Ecosystem[1]	Loss
	(percent)
Mesopotamian Marshlands, Iraq/Iran	>90
Sistan Wetlands, Iran/Afghanistan	99
Las Tablas de Daimiel, Spain	68
Everglades, United States	~50
Lake Gilli, Armenia	100

[1] Data are lacking for most wetlands in tropical Africa, Latin America, and Asia.

Source: UNEP, Alvarez-Cobelas et al., World Resources Institute, Balian et al.

around the world. South American water hyacinth chokes waterways and smothers native vegetation in Asia and Africa.[18] In the San Francisco Bay and other Pacific coast wetlands, introduced smooth (or Atlantic) cordgrass hybridizes with native grasses and is changing the plant composition and overwhelming mudflats important to threatened wildlife.[19] In the eastern United States, introduced European mute swans compete with native water birds for food and nesting habitat, and South American nutria (coypu) chew down marsh vegetation and increase erosion rates.[20] On the other side of the Atlantic, American crayfish are changing food chain dynamics in Spanish wetlands, where they have thrived following their introduction.[21]

The 1971 Convention on Wetlands, also known as the Ramsar Convention, focuses on the conservation and sustainable use of wetlands. As of December 2004, 143 countries had signed onto this treaty, together designating 1,400 wetlands as special areas warranting protection—they cover an area greater than France, Germany, and Switzerland combined (almost 123 million hectares).[22] Although it does not direct any punitive action against violators, the treaty has garnered more attention and better conservation status for many wetlands. As happened with Iran's Ramsar site in the Sistan wetlands, however, such recognition cannot by itself keep an ecosystem wet and functioning.

Global conservation agreements such as Ramsar have helped raise awareness and concern for wetlands in many countries, but enforcement of conservation laws remains lax in most areas. In addition, little is known of the status and extent of many of the world's wetlands. Without inventories to provide a baseline for action, many of the wetlands will disappear before they can be identified and sustainably managed.[23] To date, large-scale inventories have been launched only in North America and Western Europe.[24] In other areas where surveys have been conducted at all, researchers found key wetland areas to be unprotected or poorly protected, particularly in many parts of Africa, Oceania, Asia, and Central and South America.[25]

Even the United States, a country with detailed wetland protection laws and inventories, has yet to stop wetland losses within its borders, although the net loss rate has slowed. Between 1986 and 1997, 496,000 hectares (4.7 percent) of often-small, often-isolated freshwater emergent wetlands were lost, as were about 496,000 hectares (2.3 percent) of forested wetlands.[26] Coastal wetland habitats also lost ground. Net loss figures for all wetlands in the lower 48 states mask these losses because they are offset by increases in humanmade water bodies such as suburban, golf course, and aquaculture ponds.[27] While categorized as wetlands, these habitats do not provide natural goods and services to match those of naturally occurring habitats. The lower 48 states already lost an estimated 53 percent of their wetlands over the 200 years before the 1980s.[28]

Wetlands are unmatched as vital natural systems that sustain life on our planet. Without priority given to studying and protecting these resources, prospects for human health and for a large slice of the planet's biodiversity will continue to worsen.

Forest Loss Continues

Gary Gardner

Deforestation remains a serious issue globally, as many countries continue to lose more trees than they regenerate. And in countries with expanding forest area, new growth is often of lower-quality plantation forests, which are cultivated to produce harvestable wood and are less ecologically complex than natural forests.

Global forest cover stands at approximately half the original extent of 8,000 years ago.[1] Modern rates of deforestation are a matter of dispute, however, because there is no common agreement on how to measure or even define a forest. The U.N. Food and Agriculture Organization (FAO), in its *2000 Global Forest Resources Assessment* (known as FRA 2000), which has the most recent data available, reported a net loss of 9.4 million hectares of forest a year during the 1990s, an annual loss roughly the size of Portugal.[2]

LINKS p. 64

Gross forest losses were even larger, at some 14.6 million hectares, but they were offset by annual increases in natural forests and plantations of 5.2 million hectares.[3] Africa and the Caribbean had the highest rates of deforestation, each losing 0.8 percent of forested area.[4] Although FRA 2000 has been roundly criticized for using a methodology that underestimates the level of deforestation, other studies have found even lower levels of deforestation in the 1990s in some regions.[5]

Since 2000, deforestation has continued to be a concern in some of the world's major logging countries. Indonesia loses nearly 2 million hectares of forest annually, double the rate of the 1980s. The country's forest cover fell by 40 percent between 1950 and 2000, from 162 million to 98 million hectares.[6] Meanwhile, Brazil announced in 2004 that deforestation rates equaled the near-record highs of 2003.[7] More than 2.3 million hectares of Amazon forest—about half the area of Switzerland—were cut between August 2002 and 2003.[8] If global deforestation rates have been the same since 2000 as they were in the 1990s, Brazil and Indonesia account for nearly half of the world's forest losses.

Deforestation is a concern because forests are important in regulating the planet's carbon and hydrological flows and provide a host of local environmental services. Trees are essentially carbon warehouses. Their carbon is released to the atmosphere when the roots of felled trees rot in the ground and when the paper or wood products made from trees decompose in landfills. Indeed, land use changes—primarily deforestation—accounted for an estimated one third of global carbon emissions between 1850 and 1998.[9] Trees also regulate global water flows among land, the atmosphere, rivers, lakes, and oceans.

At a local level, forests provide a long list of environmental and economic services. They are home to a tremendously broad range of species, for example. Indonesia accounts for only 1.3 percent of Earth's land surface, but it has 11 percent of the world's plant species, 10 percent of the mammal species, and 16 percent of the bird species.[10] And in Brazil, deforestation is occurring at the greatest clip in areas that hold the key to species conservation in the Amazon. One of these areas has disappeared, and five others have lost half of their forest cover.[11]

Less tree cover can also reduce rainfall, since trees transpire moisture into the air, which later falls as precipitation.[12] Researchers in Colombia, one of the most water-rich countries in the world, estimate that by 2025 some 70 percent of the country's people will experience water shortages in times of drought, partly because deforestation has increased flooding and reduced the land's capacity to retain water.[13] The 2004 devastating flooding in Haiti was blamed in part on denuded hillsides, which were unable to hold water because roots that once held soil in place were no longer there.[14]

Deforestation is a complex phenomenon with many direct and underlying causes.[15] Immediate drivers include agricultural expansion, wood harvesting, and infrastructure expansion such as road building. Underlying drivers include poverty, economic growth, and other economic factors; government policies; technological advances; demographic change; and cultural factors. Other variables, such as land characteristics and soil and water profiles, along with social triggers such as war can also influence the extent of deforestation.

A 2001 analysis of 152 case studies of tropical deforestation dating back as far as 1880 challenged two prominent schools of thought about deforestation: that it typically has a single cause, such as shifting cultivation or population growth and, alternatively, that it is so complex that no clear causal patterns can be identified.[16] The analysis found several commonly occurring combinations of drivers of tropical forest loss.

Agricultural expansion was the most common explanation for deforestation in the study, appearing in 96 percent of cases.[17] But it was rarely the sole explanation. (Indeed, single-factor explanations of forest loss were found in only 6 percent of cases.)[18] Agricultural expansion was linked with wood harvesting and infrastructure expansion—especially road and railroad construction, but also settlement expansion and the establishment of mines, oil wells, and dams—in 25 percent of the analyzed cases.[19] Combinations of two of these three factors appeared in another 36 percent of cases.[20]

Regarding underlying causes, the study found that the nexus of agriculture, wood harvesting, and road building was often driven by a combination of economic, policy, institutional, and cultural factors. Deforestation caused directly by agricultural expansion and wood harvesting was often driven by new technologies. And agricultural expansion was often driven by population growth.[21]

The analysis also found that cultural factors play a larger role in deforestation than is commonly believed and that these continue to be important. In Indonesia, for example, logging is tied closely to political cronyism. In the late 1990s, President Suharto awarded logging concessions covering more than half of the country's forested area; some 45 percent of these are in the hands of just 10 companies.[22] And much of the supply of wood—an estimated 65 percent in 2000—is cut illegally.[23]

As the world has become more crowded and as wood, paper, and other forest resources are in greater demand, forest governance is changing. Already, some 22 percent of the world's forests are privately owned.[24] And community ownership now accounts for 11 percent of forests, a category that is projected to reach 40 percent by 2050.[25]

Certification schemes, which offer consumers assurances that wood products come from forests that are managed sustainably, are increasing in number worldwide, although reports of certified area differ substantially. The Forest Stewardship Council (FSC) reports that certified area—area that meets internationally recognized criteria and principles of forest stewardship—has grown more than tenfold since 1995, to some 47 million hectares in 60 countries.[26] (See Table 1 for the top 10 countries with FSC-certified forests.) In 2000, FAO counted some 80 million hectares, about 2 percent of the world's forested area, as certified.[27] The World Bank and the World Wide Fund for Nature joined forces and hoped to increase this number to 200 million hectares by 2005.[28]

Table 1. Top 10 FSC-Certified Areas, by Country

Country	Area
	(million hectares)
Sweden	10.10
Poland	6.20
United States	5.34
Canada	4.37
Brazil	2.63
Russia	2.12
Croatia	1.99
Latvia	1.69
South Africa	1.67
United Kingdom	1.21

Source: Forest Stewardship Council.

Air Pollution Still a Problem

Gary Gardner

Emissions of many air pollutants have declined or stabilized in industrial countries in recent years, the product of national regulations and international protocols over the past three decades that restrict the worst contaminants. Pollution levels are still unhealthy, however, particularly in light of new studies suggesting that the health risks from air pollution are greater than scientists believed even a decade ago. And in developing countries, especially nations undergoing rapid industrialization, most air pollutants are present at levels that are now causing significant numbers of deaths.

The World Health Organization (WHO) refers to six contaminants that are harmful to human health: carbon monoxide, lead, nitrogen dioxide, sulfur dioxide, ground-level ozone, and suspended particulate matter (usually in dust and smoke).[1] The six pollutants, whose mixture over the world's cities can vary widely, are generally the product of fossil fuel use in factories, power plants, and motor vehicles or the result of burning biomass such as forests or post-harvest crop stubble. (The WHO definition does not include carbon dioxide, which is implicated in climate change. A separate source of contaminants, indoor air pollution, is also not covered here.)

LINKS pp. 30, 56

Because data on contamination are uneven for urban areas globally, a comprehensive assessment of the quality of air worldwide is difficult. But a World Bank survey of more than 100 cities in industrial and developing countries that had data on emissions of sulfur dioxide or nitrogen dioxide found that the air in many urban areas remains unhealthy.[2] Some 29 percent of the cities listed recorded sulfur dioxide emissions (often associated with power plants) above maximum levels allowable under WHO guidelines, and 71 percent had nitrogen dioxide emissions (often associated with automobile use) that exceeded WHO maximums.[3]

In general, developing countries are less likely to meet WHO standards. Chinese cities are particularly hard hit. More than 80 percent of Chinese cities in the World Bank list had sulfur dioxide or nitrogen dioxide emissions above the WHO threshold.[4] And nearly half of the Chinese cities with excessive sulfur emissions registered levels at more than double the WHO standard.[5]

The health impacts of air pollution are more serious than was assumed through most of the twentieth century. Studies have found that the smallest particles of smoke and dust—less than 2.5 microns in size, about one fortieth the diameter of a human hair—pose the greatest risk to health.[6] This led scientists to conclude in studies in 2002 and 2004 that growing up in a city with polluted air is about as harmful to a person's health as growing up with a parent who smokes.[7]

Emerging evidence suggests that rapidly increasing rates of asthma may be linked to air pollution. Asthma appears to be correlated with high levels of ground-level ozone. In a southern California study, thousands of children in 12 communities—6 heavily polluted, 6 with relatively clean air—were monitored over five years.[8] Those active in sports in the communities with polluted air were three to four times more likely to have asthma as less active kids in communities with cleaner air.[9]

Studies in the Czech Republic and Mexico City found that the risk of infant death is doubled when pollution levels are the highest.[10] And lead, which is added as an anti-knock agent to gasoline in many countries, can damage the kidneys, nervous system, brain, and cardiovascular and reproductive systems. In children, it has been linked with reduced intelligence, lack of focus, and behavioral problems.[11]

Meanwhile, a 2000 World Bank study projected that on average 1.8 million people would die prematurely each year between 2001 and 2020 because of air pollution.[12] (See Table 1.)

Although air pollution is concentrated in cities, it can move well beyond them. Acidic lakes in Scandinavia have long been linked to pollution from factories in the United States, for example.[13] Recently, scientific attention has focused on the "Asian Brown Cloud"—a two-mile-thick collection of soot, fly ash, and sulfuric acid that has been parked over South Asia for more than a decade. The U.N. Environment

Table 1. Projected Premature Annual Deaths due to Urban Air Pollution, Total and by Economic Group or Region, 2001–2020

Region	Premature Deaths
	(thousand per year)
Established market economies	20
Former socialist economies	200
China	590
India	460
East Asia and the Pacific	150
Latin America and the Caribbean	130
South Asia	120
Middle East Crescent	90
Sub-Saharan Africa	60
World	1,810

Source: World Bank.

Programme (UNEP) reported in 2002 that this cloud had killed tens of thousands of people in the past 10 years, including 52,000 in India alone in 1995.[14] Originating from forest fires, wood-burning stoves, and a sharp increase in fossil fuel burning that has accompanied economic expansion in South Asia, the pollution has reportedly cut the amount of sunlight reaching Earth's surface by 10–15 percent.[15]

Smog has serious economic effects as well, especially in farming, where it is known to reduce crop yields. Ozone, which decreases plants' capacity to engage in photosynthesis, tends to reach its highest levels in the summer and in crop-growing regions around cities. More than half a dozen comprehensive studies in the United States and Europe since the 1980s have shown that yield reductions from ozone are economically significant.[16] A 2002 study of European farming, for example, determined that ozone was costing farmers more than 6 billion euros annually.[17]

The experience of industrial countries in tackling air pollution suggests that major advances are possible. A 1999 Princeton University review of 17 studies from five continents found a strong correlation between reductions in lead levels in gasoline and blood lead levels.[18] And when transportation policies that discouraged car use during the 1996 Olympic Games in Atlanta, Georgia, reduced vehicle-related pollutants by about 30 percent, the number of acute asthma attacks and health insurance claims fell by 40 percent, while pediatric emergency admissions to area hospitals dropped by 19 percent.[19]

Mounting evidence of the damage from air pollution and of the effectiveness of abatement policies has led to various efforts to phase out leaded gasoline and reduce sulfur levels in fuels globally. The world's leading engine manufacturers called for the elimination of lead by 2005.[20]

In 2002, a global Partnership for Clean Fuels and Vehicles was established at the World Summit on Sustainable Development.[21] UNEP, one of the partners, has spearheaded the cause in Africa in particular and announced in 2004 that more than 50 percent of the gasoline sold in sub-Saharan Africa was now lead-free—a major advance for a continent that has been slow to address the lead issue.[22]

Meanwhile, the American Academy of Pediatrics, alarmed at evidence that pollution in the United States continues to pose a serious health risk to children, called in December 2004 for stricter emissions standards for ozone, nitrogen dioxide, and particulate matter; higher fuel economy standards; the promotion of alternative fuels; and support for public transportation, carpooling, walking, and cycling.[23]

Economy and Social Features

Courtesy of USAID

Local women leading boardwalk tours through Kisite Marine National Park, Kenya

▶ Socially Responsible Investing Spreads

▶ Interest in Responsible Travel Grows

▶ Global Jobs Situation Still Poor

Socially Responsible Investing Spreads

Erik Assadourian

In 2003, investments using socially responsible criteria exceeded $2.63 trillion worldwide.[1] The United States, with the most developed socially responsible investment (SRI) market, accounted for $2.16 trillion of this total.[2] While the SRI market in Western Europe is a distant second at $413 billion, it is growing rapidly—a trend that is expected to continue.[3] Australia and Canada added another $55 billion, while SRI in Asia and emerging markets is undeveloped; each totaled less than $3 billion.[4]

Socially responsible investing, in the broadest terms, includes three major approaches: screening, shareholder advocacy, and community investment.[5] Screening can involve negative screens that exclude unacceptable industries, such as tobacco, weapons, or nuclear energy, and positive screens that select companies with superior environmental and labor records that produce safe and useful goods. Shareholder advocacy, a strategy often coupled with screening, refers to investment in a company in order to influence its decisions through shareholder resolutions. The third strategy is community investing, in which people invest in communities, often ones that are underserved by other financial services, in order to increase total capital flows to them.

SRI has proved itself to be an investment strategy not just for social reformers but for all investors. The major SRI portfolios have shown themselves to be competitive with conventional ones. For example, over the past 10 years the Domini 400 Social Index—a portfolio based on the S&P 500 that screens out 250 unacceptable corporations and adds 150 socially responsible ones—has provided an average 12.6 percent in returns each year, while the S&P 500 has given 11.9 percent.[6]

SRI has a long history, especially with religious organizations. As early as the seventeenth century, the Quakers screened out weapons companies from their investments.[7] In the 1920s, the Methodist Church in the United Kingdom used negative screens to avoid investment in "sin stocks."[8] But it was the Pax World Fund, an SRI mutual fund created in 1971, that launched the modern SRI movement.[9] By 1984, $63 billion

had been invested in SRI funds in the United States.[10]

In the 1990s, SRI started to flourish in the United States, with investments nearly doubling between 1995 and 1997 and almost doubling again two years later.[11] (See Figure 1.) While total SRI funds declined 10 percent between 2001 and 2003, the percentage of SRI funds compared with total investments under professional management remained stable at around 11 percent, since total investments declined from $20.6 trillion to $19.2 trillion.[12]

Of the $2.16 trillion invested in SRI in the United States in 2003, $1.7 trillion was in the form of screens, while $441 billion used both screens and shareholder advocacy.[13] Another $7 billion went into just shareholder advocacy, and $14 billion went to community investing.[14] While the funds invested in shareholder advocacy declined 52 percent between 2001 and 2003, the number of resolutions filed increased 15 percent.[15]

The success of shareholder resolutions is often limited, both because successfully passed resolutions are nonbinding and because a large percentage of shares are owned by non-voting institutions.[16] Indeed, the average vote in favor of resolutions that addressed social responsibility issues was just 11.4 percent in 2003 (compared with 8.7 percent in 2001).[17] The power of shareholder advocacy comes more from the pressure that corporations feel to change policies when confronted by shareholder activists. Of the 292 resolutions filed in 2002, 95 were withdrawn before voting after policy changes were agreed on with the management.[18]

Community investments, while significantly smaller than the other forms of SRI, are growing quickly, with total investments increasing 84 percent between 2001 and 2003.[19] Community development financial institutions (CDFIs), including development banks, credit unions, loan funds, and venture capital funds, finance projects that build affordable housing, create livable-wage jobs, or provide essential services such as health care.[20] Although the investments are comparatively small, the effects of community investing are impressive. A survey of 442

U.S. CDFIs found that in 2002 these institutions financed 7,800 small businesses that established or sustained 34,000 jobs, and they facilitated the building or renovation of 34,000 units of affordable housing and over 500 community facilities.[21]

Europe's SRI market is growing rapidly, with more than $413 billion currently invested in this market.[22] Of these funds, $268 billion is in screened portfolios, while $145 billion is used for shareholder advocacy.[23] The Netherlands and the United Kingdom account for 98 percent of the $413 billion, primarily from the heavy investment of pension funds.[24]

In the United Kingdom, the shift of pension funds to SRI was stimulated by a law requiring them to disclose how much consideration they give to social, environmental, and ethical issues.[25] A second law, which requires charities to ensure investments are in line with the charity's stated goals, further strengthened the SRI sector there.[26] While the SRI markets in other European countries are still undeveloped, as these countries pass similar pension legislation, which many are currently considering, SRI could become increasingly the norm across Europe.[27]

The SRI market in Asia—just $2.5 billion—is still very immature.[28] But it is growing quickly: between 2001 and 2002, the number of new funds increased 32 percent.[29] The Japanese market, 40 percent of the regional total, is poised to take off in coming years as pension funds—as in Europe—start to adopt SRI initiatives.[30]

In most developing economies, SRI virtually does not exist. While the total is estimated at about $2.7 billion, $1.5 billion comes from investors from industrial countries and 95 percent of the remainder is invested in South Africa.[31] In the short term, growth of SRI is expected to be slow because of the lack of data needed to maintain SRI initiatives, as well as strong competition with non-stock investments.[32]

In addition to pension funds, another large

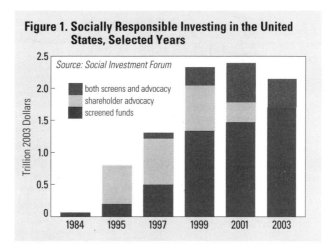

Figure 1. Socially Responsible Investing in the United States, Selected Years

Source: Social Investment Forum

both screens and advocacy
shareholder advocacy
screened funds

source of funds may start being mobilized in 2005. In April a new investment group, 3iG (the International Interfaith Investment Group), plans to start working with religious leaders around the world to invest religious organizations' funds more responsibly. According to 3iG's estimate, the central institutions of religious organizations have more than $7 trillion in assets.[33] Committing these monies to SRI could more than triple current global SRI investments and create a powerful new shareholder advocacy force. 3iG is optimistic that not only will it be able to leverage $1 trillion of these assets but that involvement by religious leaders will spark investment by regional and local chapters as well as by individual members.[34] This "cascade effect" could multiply the total benefit severalfold and trigger unprecedented growth in SRI.[35]

Interest in Responsible Travel Grows

Zoë Chafe

Tourism is "travel undertaken for pleasure," and countries around the world recorded an all-time high of some 760 million international tourism arrivals in 2004—10 percent more than in 2003.[1] The Cruise Lines International Association reported a 10-percent increase, to 7.9 million passengers worldwide, during 2004—a rate that mirrors overall tourism growth.[2]

The World Tourism Organization predicts that international tourist arrivals will reach 1.56 billion by 2020, with 46 percent of the people visiting Europe, 25 percent East Asia and the Pacific, and 18 percent the Americas.[3] The organization cites, among other things, a proliferation of low-cost airlines, independent travel, and special interest travel as trends driving the upsurge in tourism.[4]

Tourism plays a major role in the world's economy, contributing an estimated $5.49 trillion of economic activity in 2004.[5] Nearly 215 million jobs—8.1 percent of total world employment—are directly or indirectly linked to travel and tourism, while 73.7 million people work directly in the travel and tourism industry.[6]

In an effort to embrace tourism, some countries have become overly dependent on this single industry, however. At least 10 countries currently derive more than half their gross domestic product from tourism.[7] The tsunami that destroyed parts of Thailand, Sri Lanka, India, and other countries in that region in late 2004 illustrated the precarious position that overdependence on tourism can put poor countries in.

In many cases, the bulk of tourists' money does not directly benefit the area they visit. The U.N. Environment Programme estimates that when tourists travel on an all-inclusive package tour, about 80 percent of the money they have paid goes to airlines, hotels, and other international companies.[8] These businesses are often headquartered in the tourists' home country, so the host community sees little economic benefit from the visits. Even if a tourist spends money directly in a vacation spot, it is often spent on goods that were imported to meet foreign tourists' standards.[9]

Many of the other negative effects of tourism are "externalized"—environmental and social costs are not factored into the price of a tourism product, such as a package tour or airplane flight. Low-cost airlines have decreased ticket prices dramatically, for instance. Someone can now fly between Indian cities for $11, between European cities for $20, and across the entire United States for less than $100.[10] But as the number of people traveling and the frequency of trips increases, air travel contributes more to air pollution and climate change.

In an innovative move to account for this, tourists can now choose a company that will "offset" the carbon emissions produced by their flights.[11] Someone flying 2,886 kilometers round-trip from London to Rome would need to pay $17.23 to account for his or her share of carbon—about 0.5 tons—released during the flight; on a 25,659-kilometer round-trip from New York City to Johannesburg, each passenger is responsible for 3.7 tons of carbon, which costs approximately $85.47 to offset.[12] By buying credits equal to the distance of the flight taken, travelers can ensure that their money is invested in green technologies or reforestation products that will help to counter those emissions.

A growing number of tourists are also seeking an alternative to conventional "mass tourism." More than two thirds of U.S. and Australian travelers and 90 percent of British tourists consider active protection of the environment, including support of local communities, to be part of a hotel's responsibility.[13] Numerous tour operators, hotels, and tours now offer "responsible tourism" opportunities.[14] (See Table 1.) (Adventure tourism and nature-based tourism, which are niche markets within mass tourism, do not address tourists' impacts on their destinations.) Responsible tourism is about ethics and principles. Ecotourism, pro-poor tourism, geotourism, and sustainable tourism are all founded on the idea that, if done well, travel can have a positive overall impact.[15]

The increasing market demand for responsible tourism has led many businesses to adopt names suggesting they are environmentally responsible. While some are indeed examples of true ecotourism, many others are not. They may make superficial changes to their operations, encourage guests to reuse towels (a move that saves water, but that is often motivated by a desire to cut costs), or actually do nothing to improve their operations.

One way to counteract such "greenwashing" and to identify truly responsible tourism is to look for accommodations, tour operators, and other products that have been certified as responsible. Certification programs are designed to measure a company's performance against a triple bottom line (economics, environment, and social and cultural criteria), to increase industry standards, to provide consumers with information, and to create a market advantage for certified businesses.

Certification programs vary as much in how they operate as in the range of products they certify. Some programs focus on the process that businesses use to create a tourism product, while others, such as Costa Rica's Certification for Sustainable Tourism, focus on performance.[16] There is a move to standardize the tourism certifications developed independently. A Sustainable Tourism Stewardship Council, due to be launched in 2006, could help serve this function. It would unite regional initiatives such as the pilot Sustainable Tourism Certification Network of the Americas, which has developed regional baseline standards and has 62 members representing 40 organizations in 19 countries.[17]

Travelers are often so moved by their experi-

Table 1. A Typology of Tourism

Category	Definition
Adventure tourism	A form of nature-based tourism that incorporates an element of risk, higher levels of physical exertion, and the need for specialized skills.
Ecotourism	Responsible travel to natural areas that conserves the environment and improves the welfare of local people.
Geotourism	Tourism that sustains or enhances the geographical character of a place—its environment, heritage, aesthetics, and culture and the well-being of its residents.
Mass tourism	Large-scale tourism, typically associated with "sea, sand, sun" resorts and characteristics such as transnational ownership, minimal direct economic benefit to destination communities, seasonality, and package tours.
Nature-based tourism	Any form of tourism that relies primarily on the natural environment for its attractions or settings.
Pro-poor tourism	Tourism that results in increased net benefits for poor people.
Responsible tourism	Tourism that maximizes benefits to local communities, minimizes negative social or environmental impacts, and helps local people conserve fragile cultures and habitats or species.
Sustainable tourism	Tourism that meets the needs of present tourists and host regions while protecting and enhancing opportunities for the future.

Sources: Merriam-Webster Dictionary, National Geographic Traveler, The International Ecotourism Society, World Tourism Organization, Pro-Poor Tourism, Encyclopedia of Ecotourism, and Responsible travel.com.

ences that they want to contribute in some way to host communities, many of which are impoverished. A variety of tourism companies are setting up philanthropic offshoots to help tourists make meaningful donations. The Africa Foundation, created in 1992 by the safari company Conservation Corporation Africa, has raised $4 million for education, health, and income-generating programs near its lodges in East and South Africa.[18] This money has financed training for 250 teachers, university scholarships for 120 students, and the creation of 65 classrooms and 18 preschools around the company's lodges.[19] Similarly, Airline Ambassadors International, an organization with strong ties to the airline industry, has facilitated the delivery of $14 million worth of food, clothing, and medical supplies since 1996, receiving up to $500,000 per year in volunteer donations of time and aid.[20]

Global Jobs Situation Still Poor

Lisa Mastny

An estimated 2.8 billion people—roughly two fifths of the world's population—were formally employed in 2003, more than ever before, according to the International Labour Organization (ILO).[1] But this estimate masks serious challenges facing workers and families across the planet. Roughly half of the world's workers—some 1.4 billion people—struggle to survive on less than $2 a day.[2] In sub-Saharan Africa, nearly 90 percent of workers fall into this category of the "working poor," most of them unable to earn a secure income or livable wage.[3]

LINKS pp. 48, 64

Meanwhile, an estimated 186 million people were either without work or looking for a job in 2003—also a record number.[4] This represents an overall unemployment rate of 6.2 percent, up slightly from 6.1 in 2002 and up significantly from 5.6 percent in 1993.[5] (See Table 1.) The ILO attributes the overall jump in joblessness to sluggish recovery of the world economy, the ongoing conflict in Iraq, the global threat of terrorism, and the effects of diseases such as SARS on tourism in Asia. Longer-term factors include shrinking jobs in communications and information technology, widespread cuts in the manufacturing and the travel and tourism sectors, and a trend toward "informal" labor.[6]

Unemployment rates vary significantly among regions. In the Middle East and North Africa and in sub-Saharan Africa, joblessness hovers at more than 10 percent, while in East Asia, bolstered by strong economic growth in China, it stands at only 3.3 percent.[7] Over the past decade, however, the greatest increase in unemployment has occurred in countries in transition, such as Russia and Kazakhstan, which are still struggling to recover from the early 1990s.[8] Southeast Asia has also seen its unemployment rate jump from 3.9 to 6.3 percent.[9]

Joblessness in parts of Asia will no doubt rise further due to the December 2004 tsunamis, which destroyed the livelihoods of an estimated 1 million people in Indonesia and Sri Lanka alone.[10] In Sri Lanka, unemployment in the affected areas is thought to have more than doubled, to 20 percent, with the most serious losses in the vital fishing and tourism industries.[11]

Unemployment has generally fallen in industrial countries over the past decade, but it remains high in some areas. In 2003, rates neared 8 percent in Europe, reflecting a lag in the recovery of labor markets compared with gross domestic product.[12] The United States, meanwhile, is pulling itself out of the longest period of "jobless growth" in postwar history, with unemployment at around 6 percent.[13] Americans work much longer hours than Europeans—an average of 1,825 hours in 2002, compared with 1,444 hours for Germans and 1,545 for French.[14] (South Koreans worked the longest—2,447 hours.)[15] During the late 1990s, the share of the U.S. workforce putting in more than 50 hours a week grew from 15 to 20 percent, while in Europe it is well under 10 percent.[16]

Many industrial countries have seen a rising trend toward "outsourcing"—the movement of jobs to other countries, ranging from labor-intensive manufacturing positions to high-skilled work such as software design. This has contributed to some, though so far not significant, job losses back home. According to the U.S. Department of Labor, in the first three months of 2004 less than 2 percent of mass layoffs in the United States were the result of outsourcing.[17]

Many people in the developing world are not actually "jobless" per se. In the absence of state-provided unemployment insurance, social security benefits, or other social protections, they are often forced to engage in some form of economic activity, however meager. This can result in relatively low recorded unemployment in countries that are in fact very poor. At the same time, countries that are relatively well off may register high levels of unemployment, in part because people can afford not to work or to wait for more desirable jobs.[18] Absolute definitions of employment are also complicated by the growing trend toward part-time or temporary work in some countries.

Another trend in many developing countries has been a rise in informal labor. Informal workers—who engage in operations not registered or regulated by labor, health, and tax laws—are replacing traditional factory workers as the "ideal" employee in several economies.[19] In 2000, the informal sector employed more than 92 percent of India's labor force.[20] Virtually invisible, informal laborers are among the world's most marginalized groups, facing low and insecure wages, few social benefits, and little state protection.

Women account for a large share of the workers in the informal sector. In many countries, their daily activities—such as growing food or participating in the family business—are not counted as formal labor. Meanwhile, they may face economic, social, and cultural barriers that impede them from actively seeking work.[21] Female participation in the workforce is the lowest in the world in the Middle East, where cultural mores discourage women from working, though the situation is improving slowly.[22]

As women's contributions are increasingly recognized, the gap between male and female employment rates worldwide has narrowed since 1980.[23] Today, more women work than ever before—accounting for 40 percent of the world's workers.[24] Yet even where they are formally employed, women tend to have fewer training and job opportunities and face lower pay than men. Men dominate work in industry and agriculture, while women are employed in greater numbers in service-sector jobs such as nursing or social work.[25]

Over the past decade, unemployment rates among young people worldwide skyrocketed from 11.7 percent to a record 14.4 percent in 2003. According to the ILO, an estimated 88.2 million people aged 15 to 24 were without work in 2003, accounting for nearly half the world's jobless. In the developing world—home

to 85 percent of young people—unemployment in this group can be nearly four times the rate among adults.[26]

A predominance of unemployed youth can create social and political challenges.[27] Agriculture remains the single largest source of livelihood worldwide, employing 40 percent of the workforce in developing countries. But as land is degraded or subdivided, many young farmers find themselves increasingly disinherited.[28] Militias and other insurgent organizations can offer an alternative source of social mobility and self-esteem, particularly in poorer or politically repressive countries. In Uganda, large numbers of young jobless men have turned to alcohol, suicide, or violence.[29] Similarly, the lack of opportunities has been linked to ongoing instability in the Middle East, where 58 percent of the population is under the age of 25 and a quarter of working-age youth are unemployed.[30]

Despite the difficulties in many regions, the actual proportion of the "working poor" in global employment has been declining steadily, from 40.3 in 1980 to 19.7 in 2003.[31] The greatest challenge remains in sub-Saharan Africa, where the labor force continues to surge, unemployment remains high, AIDS is rampant, and economic growth has been stagnant.[32]

Table 1. Total and Youth Unemployment Rates, Worldwide and by Economic Group or Region, 1993 and 2003

Region	Total Population		Young People (Aged 15–24)	
	1993	2003	1993	2003
	(percent)		(percent)	
Middle East and North Africa	12.1	12.2	25.7	25.6
Sub-Saharan Africa	11.0	10.9	21.9	21.0
Transition Economies	6.3	9.2	14.9	18.6
Latin America and the Caribbean	6.9	8.0	12.4	16.6
Industrial Economies	8.0	6.8	15.4	13.4
Southeast Asia	3.9	6.3	8.8	16.4
South Asia	4.8	4.8	12.8	13.9
East Asia	2.4	3.3	4.8	7.0
World Total	5.6	6.2	11.7	14.4

Source: ILO.

Governance Features

A child scavenging on "Smokey Mountain," Manila, Philippines

▶ Global Public Policy Cooperation Grows

▶ Greater Effort Needed to Achieve the MDGs

Global Public Policy Cooperation Grows

Molly Aeck

In recent years, a new breed of governance has emerged in response to the rapidly changing conditions of a globalizing world. Known to some people as global public policy networks, or GPPNs, these innovative groups bring various partners—governments, international organizations, the private sector, and nongovernmental organizations (NGOs)—together under the umbrella of a cohesive working group to address shared international challenges.[1] Since the early 1990s, approximately 50–60 GPPNs have emerged, focusing on everything from crime and malaria to fisheries and agriculture.[2] (See Table 1.)

The growth of these networks is linked to the rise of new actors on the world stage. As countries become increasingly connected through the integration of markets, for instance, corporations play a more prominent global role.[3] The number of multinational enterprises grew from just a few hundred in the early 1970s to well over a thousand in 1990.[4] Meanwhile, the advent of the Internet and other new communications tools has spurred a new engine of participatory democracy through global activism. NGOs have grown in number from fewer than 5,000 in 1975 to more than 25,000 in 2000—emerging as an additional major force in global politics.[5]

In the past, governments and international organizations such as the United Nations assumed primary responsibility for addressing cross-border challenges, from the foreign repercussions of financial crises to global climate change. But even the more inclusive instruments of such international policymaking—international treaties and cooperation through shared institutions and agencies—are often no longer sufficient. For this reason, governments are increasingly forming partnerships and networks with new and wider constituencies, including corporations and NGOs, to develop and implement policy.

Although GPPNs only became popular in the 1990s, one early network was the Consultative Group on International Agricultural Research, created in 1971 to increase sustainable food production.[6] A more recent example is the Kimberley Process Certification Scheme, a coopera-

tive arrangement of diamond companies, governments, and civil society organizations that certifies that exported diamonds are not "conflict diamonds"—rough gems sold to fund civil conflict.[7] And the African Stockpiles Program (ASP), formed in 2001, offers a seat at the policymaking table for civil society and the private sector.[8] By involving the plant science, agriculture, and pesticide industries in a dialogue with civil society and governments, the ASP is able to effectively target and remove stockpiled pesticides and pesticide-contaminated waste in Africa.

Global public policy networks often emerge when policymakers lack the resources or support to address complex policy issues that require far-reaching consensus.[9] Through the International Forum on Forests, for instance, stakeholders jointly develop proposals that provide governments, international organizations, and private-sector entities with guidance on how to further develop, implement, and coordinate national and international policies on sustainable forest management.[10] Similarly, the Global Water Partnership is a mechanism for alliance building and information exchange among stakeholders working toward integrated water resources management.[11]

Because of their decentralized, inclusive, nonhierarchical structure and their transparent operation, GPPNs are well placed to engage in broad policy negotiations. In the absence of such networks, international cooperation is often honored more in word that in deed, as is the case when finance ministers from industrial countries meet regularly to evaluate policy objectives but do little to adjust for common interests. The dense networks of interaction embedded in GPPNs, in contrast, reduce incentives for opportunism and misconduct by allowing groups that may traditionally disagree to find common ground on various mutual interests.[12]

A GPPN's structure depends largely on the purpose it serves. Typically, the network's creators initially outline its legal status and membership codes, as well as agree on governing, management, or technical advisory bodies.

Table 1. Timeline of Selected Global Public Policy Networks

Year Created	Network Name	Function	Details
1971	Consultative Group on International Agricultural Research	Generation and dissemination of knowledge	Broadens support for research in food supplies and sustainable agriculture.
1992	International Campaign to Ban Landmines	Advocacy and agenda-setting	Calls for a ban on antipersonnel landmines and greater international support for mine clearance and victim assistance.
1997	Global Reporting Initiative	Standard-setting	Develops and disseminates global sustainability reporting guidelines for companies.
1998	World Commission on Dams	Standard-setting	Conducted a comprehensive global review of the impacts of large dams and negotiated criteria for new projects. In 2001, this led to creation of the Dams and Development Project, translating recommendations into on-the-ground action.
1999	Medicines for Malaria Venture	Finance	Brings public, private, and philanthropic partners together to fund the discovery, development, and registration of new medicines for the treatment and prevention of malaria.
2000	Millennium Ecosystem Assessment	Generation and dissemination of knowledge	Analyzing scientific information on the consequences of ecosystem change for human well-being and options for responding to those changes.
2002	Kimberly Process Certification Scheme	Standard-setting	Certifies that sales of rough diamonds are not used by rebel groups to finance armed conflict.
2005	Renewable Energy Policy Network	Advocacy and agenda-setting	Provides international leadership on policy initiatives spurring the development of renewable energy.

Source: Web site for each network.

In the early stages, an established organization may serve as the secretariat until a more permanent governing body and constituency is created—as the World Resources Institute did for the Millennium Ecosystem Assessment.[13]

GPPNs are characteristically open to any group or individual interested in supporting its goals, and any government can join through its contributions to the process. Members often represent very different sectors and levels of governance. This flexibility and open membership enables networks to address international problems quickly.[14] The Millennium Ecosystem Assessment's ability to network simultaneously at the community, watershed, national, regional, and global levels, for instance, has contributed greatly to its success.[15]

Every member brings different resources to the table, and even unlikely partners can jointly guide the development, implementation, and coordination of policies. This cooperation may bring new issues to the international agenda or may increase the prominence of existing issues by articulating clear and focused goals. Ultimately, a network might focus its efforts on one particular goal, such as revitalizing weak treaties, coordinating research initiatives, negotiating guidelines to address urgent policy issues, or improving market design.

Inevitably, conflicts arise among network members. The most successful GPPNs, however, are powerful enough to provide the mutual support, cooperation, trust, and institutional effectiveness to overcome these challenges for

Greater Effort Needed to Achieve the MDGs

Hilary French

In September 2000, world leaders present at the U.N. Millennium Assembly adopted an ambitious set of Millennium Development Goals (MDGs) to be achieved by 2015.[1] (See Box 1.) Specific targets within the MDGs include cutting poverty and hunger rates in half, reducing child mortality by two thirds, and halving the proportion of people lacking access to clean drinking water and adequate sanitation.

pp. 64, 68, 102

Progress toward achieving the MDGs will be reviewed at another summit-level meeting at the U.N. General Assembly in New York in September 2005—a gathering that U.N. Secretary-General Kofi Annan has called "an event of decisive importance… that may offer us our best—perhaps our only—chance to ensure a safer, more just, and more prosperous world in this new century."[2]

As will no doubt be confirmed at the upcoming Summit, the prognosis for achieving the MDGs is mixed. On the encouraging side, some countries and regions have made significant gains in reducing poverty rates over the last decade, and the world as a whole is generally believed to be on track for meeting the targets set for poverty reduction and clean drinking water. But the situation is less hopeful for the other goals and targets, including those on hunger, primary education, child mortality, and access to sanitation.[3] (See Table 1.) According to the World Bank, less than one fifth of all countries are currently on target to reduce child and maternal mortality and provide access to water and sanitation, while even fewer are on course to contain HIV, malaria, and other major diseases.[4]

Experts agree that most governments are not currently making enough effort to achieve the MDGs. On the other hand, evidence abounds that when governments do set the achievement of certain goals as a priority, they can rapidly register great success.[5]

Among other steps, achieving the MDGs will require increased public investment. In a January 2005 report to the Secretary-General, the UN Millennium Project cited a long list of "quick wins"—high-impact interventions that can be implemented rapidly. Among the initiatives highlighted by the report are programs to distribute malaria bed nets and effective antimalaria medication to children, end user fees for primary schools and essential health services, expand school meal programs, and provide fertilizer to African farmers.[6]

Some countries have already recognized the need to shift domestic spending priorities to help finance these sorts of programs, in part by reducing military expenditures. In 2003, for example, Brazil delayed the purchase of $760 million worth of jet fighters and cut its military budget by 4 percent in order to finance an ambitious anti-hunger program.[7] And Costa Rica, by having no military for the past 50 years, has been able to devote a much larger portion of its budget to social spending—with impressive results. With a similar gross domestic product (GDP) per capita as Latin America as a whole, Costa Rica has the highest life expectancy rate and one of the highest literacy rates in the entire region.[8]

But many countries will need more funding for MDG-related initiatives than they can generate internally. The World Health Organization estimates, for example, that to sustain a public health system, a minimum of $30–40 per person is necessary.[9] In the world's poorest countries, where GDP per capita is typically in the low hundreds, even this rather modest level of spending will be impossible without outside investment.[10]

As the final MDG makes clear, a concerted effort from industrial countries and global institutions will thus be essential—both through additional development aid and through broader

---

Box 1. Millennium Development Goals

- Eradicate extreme poverty and hunger
- Achieve universal primary education
- Promote gender equality and empower women
- Reduce child mortality
- Improve maternal health
- Combat HIV/AIDS, malaria, and other diseases
- Ensure environmental sustainability
- Develop a global partnership for development

Source: United Nations.

Table 1. Global and Regional Progress in Achieving Selected MDG Targets by 2015

Region	Halving Poverty	Halving Hunger	Primary Education for All	Reduce Child Mortality by Two Thirds	Halve Share Without Access to Safe Drinking Water	Halve Share Without Access to Sanitation
Arab States	achieved	reversal	on track	lagging	n. a.	n. a.
Central/Eastern Europe and CIS	reversal	n. a.	achieved	lagging	achieved	n. a.
East Asia/ Pacific	achieved	on track	achieved	lagging	lagging	lagging
Latin America/ Caribbean	lagging	on track	achieved	on track	on track	lagging
South Asia	on track	lagging	lagging	lagging	on track	lagging
Sub-Saharan Africa	reversal	reversal	lagging	lagging	lagging	reversal
WORLD	on track	lagging	lagging	lagging	on track	lagging

Source: UNDP.

economic initiatives such as increased debt relief and fairer trade. In 2003, donor countries gave $68 billion in official development assistance, or just 0.25 percent of their gross national incomes, far short of the 0.7 percent of national income goal that was initially adopted at the 1970 General Assembly and broadly reaffirmed in 2002 at major international conferences in Monterrey and Johannesburg.[11] Only five countries have met the 0.7 percent target so far—Denmark, Luxembourg, the Netherlands, Norway, and Sweden.[12] If all donors were to follow their lead, annual development aid would surpass the $195 billion in funding that the UN Millennium Project estimates will be needed to achieve the MDGs.[13]

In addition to increasing overall spending, donor countries will also have to do better at targeting the aid they currently provide. In 2001, for instance, more than a fifth of all aid was conditioned on purchasing goods and services from the donor country, while less than a third went to improving basic health, sanitation, and education services.[14] Aid is also disbursed more often on political than objective social criteria. Spending on reconstruction in Iraq, for instance, now dominates overall U.S. aid spending, at $18.44 billion in 2004.[15] In contrast, all other U.S. aid spending added up to $20.67 billion that year, and more than a quarter of the funds went to just four countries—Israel, Egypt, Colombia, and Jordan—none of which are among the poorest in the world.[16]

Although political leadership is essential for moving the world closer to meeting the Millennium Development Goals, it is sobering to note that even if these targets are in fact achieved on schedule in 2015, there will still be 400 million people who are undernourished, 600 million who live on less than $1 per day, and 1.2 billion without access to improved sanitation.[17] And the world is not currently on track to meet most of the goals. To do so, governments of both the North and the South will need to make strong commitments—and then live up to them.

Notes

GRAIN HARVEST AND HUNGER BOTH GROW
(pages 22–23)

1. U.N. Food and Agriculture Organization (FAO), *FAOSTAT Statistical Database*, at apps.fao.org, updated 20 December 2004.
2. Ibid.
3. Ibid.; U.S. Bureau of the Census, *International Data Base*, electronic database, Suitland, MD, updated 30 September 2004.
4. FAO, op. cit. note 1.
5. Ibid.
6. Ibid.
7. Ibid.; FAO, *Food Outlook* (Rome: December 2004), p. 3.
8. FAO, op. cit. note 7.
9. U.S. Department of Agriculture (USDA), *Wheat: World Situation and Outlook* (Washington, DC: November 2004).
10. United Nations, "United Nations Launches International Year of Rice: 'Symbol of Cultural Identity and Global Unity'," press release (New York: 31 October 2003).
11. FAO, "Rice and Human Nutrition," International Year of Rice 2004 Fact Sheet No. 3 (Rome: 2004).
12. Ibid.
13. Rhea Sandique-Carlos, "OUTLOOK 05: Global Rice Prices May Rise On Tighter Supply," *Dow Jones*, 6 January 2005; FAO, op. cit. note 7.
14. USDA, *Rice: World Situation and Outlook* (Washington, DC: November 2004).
15. Worldwatch calculation based on ibid.
16. FAO, op. cit. note 7.
17. Ibid.
18. Ibid.
19. FAO, op. cit. note 1.
20. Worldwatch calculation from ibid.
21. Ibid.
22. Ibid.
23. Fertilizer from International Fertilizer Industry Association, "Summary Report: World Agriculture and Fertilizer Demand, Global Fertilizer Supply and Trade 2004–2005," presented at 30th IFA Enlarged Council Meeting, Santiago, Chile, 1–3 December 2004; irrigation from FAO, op. cit. note 2.
24. FAO, *The State of Food Insecurity in the World* (Rome: 2004), pp. 4–5, 34.
25. Ibid., pp. 6, 34.
26. Ibid.
27. Ibid., p. 4.

MEAT PRODUCTION AND CONSUMPTION RISE
(pages 24–25)

1. U.N. Food and Agriculture Organization (FAO), *FAOSTAT Statistical Database*, at apps.fao.org, updated 20 December 2004; FAO, "Meat and Meat Products," *Food Outlook No. 4*, December 2004.
2. FAO, *FAOSTAT Statistical Database*, op. cit. note 1.
3. FAO, *Food Outlook*, op. cit. note 1.
4. Christopher Delgado, Mark Rosegrant, and Nikolas Wada, "Meating and Milking Global Demand: Stakes for Small-Scale Farmers in Developing Countries," in A. G. Brown, ed., *The Livestock Revolution: A Pathway from Poverty?* Record of a conference conducted by the ATSE Crawford Fund, Canberra, 13 August 2003 (Parkville, Vic., Australia: The ATSE Crawford Fund, 2003).
5. Ibid.; Christopher Delgado et al., *Livestock to 2020: The Next Food Revolution* (Washington, DC: International Food Policy Research Institute, 1999).
6. Delgado, Rosegrant, and Wada, op. cit. note 4.
7. FAO, *Food Outlook,* op. cit. note 1.
8. Ibid; World Health Organization, "Cumulative Number of Confirmed Cases of Avian Influenza A/(H5N1) Since January 28, 2004," Geneva, 2 February 2005; Keith Bradsher and Lawrence K. Altman, "A War and a Mystery: Confronting Avian Flu," *New York Times*, 12 October 2004.
9. FAO, *Food Outlook*, op. cit. note 1.
10. Ibid.
11. Ibid.

12. Ibid.
13. Ibid.
14. Ibid.
15. Ibid.
16. Ibid.
17. Cees de Haan et al., "Livestock and the Environment: Finding a Balance," Report of a Study Coordinated by FAO, U.S. Agency for International Development, and World Bank (Brussels: 1997), p. 53.
18. Erik Millstone and Tim Lang, *The Penguin Atlas of Food: Who Eats What, Where, and Why* (London: Penguin Books, 2003), p. 35.
19. Ibid., p. 62.
20. Ibid.
21. Margaret Mellon, Charles Benbrook, and Karen Lutz Benbrook, *Hogging It! Estimates of Antimicrobial Abuse in Livestock* (Washington, DC: Union of Concerned Scientists, 2001).
22. David Barboza with Sherri Day, "McDonald's Seeking Cut in Antibiotics in Its Meat," *New York Times*, 20 June 2003; "Fast Food, Not Fast Antibiotics," *New York Times*, 22 June 2003.
23. Whole Foods Market, "Whole Foods Market Establishes Foundation to Help Achieve More Compassionate Treatment of Farm Animals," press release (Austin, TX: 14 December 2004).
24. I. Koizumi et al., "Studies on the Fatty Acid Composition of Intramuscular Lipids of Cattle, Pigs, and Birds," *Journal of Nutritional Science Vitaminol* (Tokyo), vol. 37, no. 6 (1991), pp. 545–54; Jo Robinson, *Why Grassfed is Best! The Surprising Benefits of Grassfed Meat, Eggs, and Dairy Products* (Vashon, WA: Vashon Island Press, 2000), pp. 12–15.

AQUACULTURE PUSHES FISH HARVEST HIGHER (pages 26–27)

1. U.N. Food and Agriculture Organization (FAO), *FAOSTAT Statistical Database*, at apps.fao.org, updated 20 December 2004.
2. Ibid.
3. Ibid.; U.S. Bureau of the Census, *International Data Base*, electronic database, Suitland, MD, updated 30 September 2004.
4. FAO, op. cit. note 1.
5. Ibid.
6. Ibid.
7. Ibid.
8. Ibid.
9. Ibid.

10. FAO, *State of the World's Fisheries 2002* (Rome: 2002).
11. Christopher L. Delgado et al., *Outlook for Fish to 2020: Meeting Global Demand* (Washington, DC: International Food Policy Research Institute (IFPRI), 2003), pp. 4–7.
12. FAO, op. cit. note 1.
13. Ibid.; Meryl Williams, *The Transition in the Contribution of Living Aquatic Resources to Food Security*, Food, Agriculture, and the Environment Discussion Paper 13 (Washington, DC: IFPRI, 1996).
14. Stefania Vannuccini, *Overview of Fish Production, Utilization, Consumption and Trade, Based on 2002 Data* (Rome: FAO, 2004), p. 3.
15. Ibid.
16. Ibid., p. 17.
17. FAO, op. cit. note 10; WorldFish Center, *Fish: An Issue for Everyone—A Concept Paper for Fish for All* (Penang, Malaysia: 2002).
18. WorldFish Center, op. cit. note 17.
19. FAO, op. cit. note 10; WorldFish Center, op. cit. note 17.
20. Ransom A. Myers and Boris Worm, "Rapid Worldwide Depletion of Predatory Fish Communities," *Nature*, 15 May 2003, pp. 280–83.
21. WorldFish Center and IFPRI, "Sea of Change in Fisheries," press release (Penang, Malaysia: 3 October 2003).
22. Aquaculture statistics from Sara Montanaro, Fishery Information, Data and Statistics Unit, FAO, e-mail to author, 7 January 2005.
23. Steven Hedlund, *SeaFood Business*, e-mail to author, 1 February 2005.
24. Jessica Wenban-Smith, Marine Stewardship Council, e-mail to author, 12 January 2005.
25. Census of Marine Life, "Making Ocean Life Count," press release (Washington, DC: 23 November 2004).

FOSSIL FUEL USE SURGES (pages 30–31)

1. International Energy Agency (IEA), *Oil Market Report*, 18 January 2005; figure includes historical data from BP, *Statistical Review of World Energy* (London: 2004).
2. IEA, op. cit. note 1; WTRG Economics, at www.wtrg.com/daily/crudeoilprice.html.
3. Author's estimate, based on Nao Nakanishi, "China Fuels Surge in Coal Demand, Prices," *Reuters*, 10 December 2004, on Energy Information Administration, *Quarterly Coal Report: July-September 2004* (Washington, DC: U.S. Department of Energy, 20

December 2004), and on IEA, *Monthly Natural Gas Survey*, October 2004.

4. Matt Pottinger, Steve Stecklow, and John J. Fialka, "Invisible Export—A Hidden Cost of China's Growth: Mercury Migration," *Wall Street Journal*, 17 December 2004.

5. IEA, op. cit. note 1.

6. Ibid.

7. David J. Lunch, "Voracious Growth Puts China in a Power Crunch," *USA Today*, 2 September 2004.

8. IEA, op. cit. note 1.

9. WTRG Economics, op. cit. note 2.

10. IEA, *World Energy Outlook 2004* (Paris: 2004), p. 81.

11. Kjell Aleklett, Uppsala University, Sweden, "The Peak and Decline of World Production of Oil," PowerPoint presentation, Asia Pacific Energy Conference, Osaka, Japan, September 2004.

12. PFC Energy, "PFC Energy's Global Crude Oil and Natural Gas Liquids Supply Forecast," PowerPoint presentation, Washington, DC, September 2004.

13. Energy Information Agency, U.S. Department of Energy, at www.eia.doe.gov/emeu/aer/petro.html, viewed 22 October 2004.

14. IEA, op. cit. note 1.

15. Ibid.

16. Matthew R. Simmons, "The Saudi Arabian Oil Miracle," PowerPoint presentation at the Center for Strategic and International Studies, Washington, DC, 24 February 2003.

17. Ibid.

18. PFC Energy, op. cit. note 12.

19. Ibid.

NUCLEAR POWER RISES ONCE MORE (pages 32–33)

1. Installed nuclear energy capacity is defined as reactors connected to the grid as of 31 December 2004 and is based on Worldwatch Institute database complied from statistics from the International Atomic Energy Agency (IAEA) and press reports primarily from *Associated Press*, *Reuters*, and *World Nuclear Association (WNA) News Briefing*, and from Web sites.

2. Worldwatch Institute database, op. cit. note 1.

3. Ibid.

4. Ibid.

5. Ibid.; "India: Construction of a 500 MWe Prototype Fast Breeder Reactor," *WNA News Briefing*, 1 September 2004.

6. Worldwatch Institute database, op. cit. note 1.

7. "US: The Department of Energy (DOE) Announced Initial Funding," *WNA News Briefing*, 9 November 2004.

8. Framatome ANP, "The AREVA and Siemens Consortium Is Awarded by TVO a Contract to Build an EPR Nuclear Power Plant," press release (Paris: 18 December 2003).

9. "France: Electricite de France Will Construct the First," *WNA News Briefing*, 26 October 2004.

10. "UK: The Chapelcross Nuclear Power Plant Shut Down," *WNA News Briefing*, 29 June 2004.

11. "The Government Confirmed its Decision to Close Barseback-2," *WNA News Briefing*, 22 December 2004.

12. "Lithuania Shuts Down Atomic Unit," *BBC News*, 31 December 2004.

13. "Romania Has Abandoned Its Approach to Complete Cernavoda-3," *WNA News Briefing*, 22 December 2004; "Romania: Atomic Energy of Canada Has Initiated Commissioning," *WNA News Briefing*, 19 October 2004.

14. "Russia: The Kalinin-3 Nuclear Power Reactor," *WNA News Briefing*, 22 December 2005; "Ukraine: Rovno-4 Nuclear Power Reactor Connected to the Grid," *WNA News Briefing*, 12 October 2004; "Ukraine: Khmelnitski-2 Nuclear Power Reactor Connected to the National Grid," *WNA News Briefing*, 10 August 2004; start dates from IAEA, "Nuclear Power Reactors in the World," April 1994.

15. Vladimir Slivyak, Co-Chairman, ECODEFENSE! and Director, Anti-Nuclear Campaign, Socio-Economic Union, Nizhegorodskaya, Russia, e-mail to author, 11 January 2005.

16. Ibid.

17. Elaine Lies, "Four Die in Steam Leak at Japan Nuclear Plant," *Reuters*, 9 August 2004.

18. "Chubu Electric Starts Japan's Biggest Reactor in Shizuoka Pref.," *Associated Press*, 18 January 2005.

19. "South Korea: The Ulchin-5 Nuclear Power Reactor Began Commercial Operation," *WNA News Briefing*, 9 November 2004; Worldwatch Institute database, op. cit. note 1.

20. "South Korea Will Move Ahead with Plans to Construct 10," *WNA News Briefing*, 30 March 2004.

21. "South Korea Enriched Nuclear Material," *WNA News Briefing*, 7 September 2004.

22. "North Korea: The Board of the Korea Energy Development Organization," *WNA News Briefing*, 19 October 2004.

23. "China Accelerates Nuclear Energy Development,"

Xinhua English Newswire, 26 September 2004.

24. "China Plans to Build PFR Nuclear Power Stations by about 2020," *Xinhua English Newswire*, 18 January 2005.

25. Worldwatch Institute database, op. cit. note 1.

26. "World Nuclear Power Reactors 2003–2005 and Uranium Requirements," World Nuclear Association (London: 1 January 2005).

GLOBAL WIND GROWTH CONTINUES
(pages 34–35)

1. Estimate for 2004 by Worldwatch with data from Birger Madsen, e-mail to author, 26 January 2005, from BTM Consult, *World Market Update 2003* (Ringkøbing, Denmark: March 2004), pp. 4–5, from American Wind Energy Association (AWEA), "U.S. Wind Industry Continues Expansion of Clean, Domestic Energy Source," press release (Washington, DC: 27 January 2005), and from European Wind Energy Association (EWEA), "Wind Power Installed in Europe by End of 2004 (cumulative)" (European map), at www.ewea.org, January 2005; European households estimated by Worldwatch with data from EWEA and AWEA, "Global Wind Power Growth Continues to Strengthen: Record 8 Billion Wind Power Installed in 2003," press release (Brussels and Washington, DC: 10 March 2004).

2. Estimated by Worldwatch with data from Madsen, op. cit. note 1, AWEA, op. cit. note 1, and EWEA, op. cit. note 1.

3. Ibid.

4. Worldwatch estimate of 65 countries based on data from BTM Consult, op. cit. note 1, and on EWEA, op. cit. note 1; Europe's share estimated by Worldwatch with data from Madsen, op. cit. note 1, from BTM Consult, op. cit. note 1, from AWEA, op. cit. note 1, and from EWEA, op. cit. note 1.

5. European data from EWEA, "Wind Power Continues to Grow in 2004 in the EU, But Faces Constraints of Grid and Administrative Barriers," press release (Brussels: 27 January 2005), and from EWEA, op. cit. note 1.

6. EWEA, op. cit. note 1.

7. Spain's total capacity from EWEA, op. cit. note 1; share of electricity demand and possible new targets from Peter Korneffel, "Seven Fat Years," *New Energy*, December 2004, p. 37.

8. Korneffel, op. cit. note 7.

9. BTM Consult, op. cit. note 1; data from EWEA, op. cit. note 1.

10. "Domestic German Wind Power Dithers," RenewableEnergyAccess.com, 1 December 2004.

11. Share in 2004 estimated by Worldwatch with data from German Wind Energy Association and VDMA Power Systems, "Domestic Sales Down, Exports Up for German-made Wind Energy Turbines," press release (Frankfurt and Osnabrück: 20 October 2004); shares in four states from ibid.; 2001 from "German 2002 Wind Power Market Up 22 Pct," *Reuters*, 24 January 2003.

12. Estimated by Worldwatch with data from Madsen, op. cit. note 1, from BTM Consult, op. cit. note 1, from AWEA, op. cit. note 1, and from EWEA, op. cit. note 1.

13. AWEA, op. cit. note 1.

14. "US Wind Boom Creates 150,000 New Jobs," *New Energy*, December 2004, p. 12.

15. AWEA, op. cit. note 1.

16. EWEA, op. cit. note 1.

17. Ibid.

18. "Second Largest UK Offshore Wind Farm Completed," RenewableEnergyAccess.com, 15 December 2004.

19. Onshore from "Wind Powers Ahead: Record Year for UK Wind as Industry Survey Predicts £7 Billion of New Investment by 2010," press release (London: British Wind Energy Association (BWEA), 21 November 2004); offshore from BWEA, *Annual Review 2004* (London: 2004), p. 3.

20. EWEA, op. cit. note 1; Madsen, op. cit. note 1.

21. "Second Largest UK Offshore Wind Farm Completed," op. cit. note 18.

22. Corin Millais, "Security from the Sea," RenewableEnergyAccess.com, 13 December 2004.

23. Madsen, op. cit. note 1.

24. "India's Wind Power Potential Has Room for Growth," RenewableEnergyAccess.com, 28 December 2004.

25. Estimated by Worldwatch with data from Madsen, op. cit. note 1.

26. Hanne May, "Wind Over the Wall," *New Energy*, December 2004, p. 13; Intergovernmental Conference for Renewable Energies, Bonn, Germany, 1–4 June 2004.

27. David Milborrow, "Goodbye Gas and Squaring up to Coal," *Windpower Monthly*, January 2005, p. 31.

28. "Three New Giant Turbines," *New Energy*, December 2004, p. 7.

29. Sales in 2004 estimated by Worldwatch with data

from EWEA and AWEA, op. cit. note 1, from Madsen, op. cit. note 1, from BTM Consult, op. cit. note 1, from AWEA, op. cit. note 1, and from EWEA, op. cit. note 1; 2012 projection from Joel Makower, Ron Pernik, and Clint Wilder, "Clean Energy Trends 2003," Clean Edge Inc., February 2003, at www.cleanedge.com/reports/trends2003.pdf.

30. Global figure is Worldwatch estimate based on Andreas Wagner, GE Wind Energy and EWEA, e-mail to author, 18 September 2002, and on EWEA and Greenpeace, *Wind Force 12* (Brussels: May 2004), p. 10; Germany from German Wind Energy Association, "Jobs in the Wind Energy Sector: Chance for the Future," 2004, at www.wind-energie.de.

SOLAR ENERGY MARKETS BOOMING (pages 36–37)

1. Paul Maycock, PV Energy Systems, e-mail to author, 27 January 2005.
2. Estimated by Worldwatch with data provided by Maycock, op. cit. note 1, and in Paul Maycock, *PV News*, various issues.
3. Estimated by Worldwatch with data provided by Maycock, op. cit. note 1, and in Paul Maycock, *PV News*, various issues. Cumulative production does not represent cumulative global installations, as all the capacity produced over the years is not still in operation.
4. Maycock, op. cit. note 1.
5. Less than 1 percent calculated by Worldwatch with data from International Energy Agency (IEA), *Key World Energy Statistics 2004* (Paris: 2004), p. 24, and from European Photovoltaic Industry Association and Greenpeace, *Solar Generation* (Brussels and Amsterdam: October 2004), p. 31; for future potential see, for example, ibid., pp. 6–7.
6. Jobs estimate based on data for Japan, Germany, and United States from Photovoltaic Power Systems Programme, *National Status Report 2003—Industry and Growth* (Paris: IEA, 2004); today's market from Mantik Kusjanto and Anneli Palmen, "Germany's SolarWorld Seeks Place in the Sun," *Reuters*, 13 January 2005.
7. Kusjanto and Palmen, op. cit. note 5.
8. Calculated by Worldwatch with data from Maycock, op. cit. note 1.
9. Paul Maycock, *PV News*, December 2004, p. 1.
10. Ibid.
11. Calculated by Worldwatch with data from Maycock, op. cit. note 1.

12. Data for 2004 from Kusjanto and Palmen, op. cit. note 5; 2003 additions from Photovoltaic Power Systems Programme, *In Brief: Germany* (Paris: IEA, 2004).
13. Robert Collier, "Germany Shines a Beam on the Future of Energy: Nation Gambles on Amped-up Push for Renewable Power," *San Francisco Gate*, 20 December 2004.
14. Calculated by Worldwatch with data from Maycock, op. cit. note 1.
15. Estimate based on data from the Photovoltaic Power Systems Programmeof IEA and from Paul Maycock, PV Energy Systems, discussion with author, 8 March 2004.
16. California Energy Commission, "Grid-connected PV Capacity (kW) Installed in California," updated 31 December 2004, at www.energy.ca.gov/renewables/emerging_renewables/2005-01-18_GRID_PV.PDF.
17. Price reduction is for systems 2–5 kilowatts peak; Photovoltaic Power Systems Programme, *Japan: National Status Report 2003—Industry and Growth* (Paris: IEA, 2004).
18. Joel Makower, Ron Pernick, and Andrew Friendly (Solar Catalyst Group), *Solar Opportunity Assessment Report* (Washington, DC: Clean Edge, Inc., and Co-op America Foundation, 2003), p. 66; Maycock, op. cit. note 1.
19. Maycock, *PV News*, January 2005, p. 1.
20. Japan from Paul Maycock, discussion with author, 23 February 2004; California from IEA, *Renewables for Power Generation: Status and Prospects* (Paris: Organisation for Economic Co-operation and Development, 2003), p. 24.
21. Georgia Institute of Technology, "Tech Developing Efficient Organic Solar Cell," press release (Atlanta, GA: 13 December 2004); "New Thin Solar Panel Could be Woven into Clothing, Says Canadian Prof," *CBC News*, 10 January 2005.
22. Estimated by Worldwatch with data from Werner Weiss, Arbeitsgameinschaft Erneuerbare Energie (AEE INTEC), Gleisdorf, Austria, e-mail to author, 23 January 2005, from Werner Weiss, Irene Bergmann, and Gerhard Faninger, *Solar Heating Worldwide: Markets and Contribution to the Energy Supply 2001* (Paris: Solar Heating and Cooling Programme, IEA, 2004), and from Eric Martinot, Worldwatch Institute, conversation with author, 30 July 2004.
23. Data for 2004 data from Weiss, op. cit. note 21.

24. Werner Weiss and Dagmar Jaehnig, AEE INTEC, Austria, "What is a Square Meter in Terms of Power?" provided by Werner Weiss, e-mail to author, 14 October 2004; "Solar Thermal Capacity Three Times Higher than Wind," *Refocus*, 17 November 2004.

25. Households worldwide estimated by Worldwatch with data from Weiss and Jaehnig, op. cit. note 23, from Eric Martinot, Worldwatch Institute, e-mail to author, 5 November 2004, and from Martinot, op. cit. note 21; shares calculated with data from Weiss, op. cit. note 21, and from Martinot, op. cit. note 21.

26. World leader and 2003 capacity from Li Hua, "From Quantity to Quality: How China's Maturing Solar Thermal Industry Will Need to Face Up to Market Challenges," *Renewable Energy World*, January-February 2005, p. 56; share from Martinot, op. cit. note 21.

27. Targets from Hua, op. cit. note 25, p. 57.

28. Martinot, op. cit. note 24.

29. "Hawaii Reaches Solar Thermal Milestone," SolarAccess.com, 12 June 2003.

30. Scott Sklar, "Selecting a Solar Heating System," *Solar Today*, September/October 2004, pp. 42–45.

BIOFUEL USE GROWING RAPIDLY (pages 38–39)

1. Lew Fulton et al., *Biofuels for Transport* (Paris: International Energy Agency, 2004), citing data from F. O. Lichts, "World Ethanol and Fuels Report" (2003).

2. Ibid.

3. Ibid.

4. James Kilner, "High Oil Prices Bolster Global Ethanol Market," *Reuters*, 11 January 2005; Fulton et al., op. cit. note 1.

5. Fulton et al., op. cit. note 1, p. 27.

6. Ibid.

7. National Biodiesel Board, at www.biodiesel.org.

8. David Cullen, "Biofuels Corp. Says UK Biodiesel Project On Track," *Reuters*, 14 January 2005.

9. "South Korea to Promote Bio Fuels to Reduce Oil Demand," *Reuters*, 1 January 2005; "Madagascar to Provide Biodiesel Production Boost," RenewableEnergyAccess.com, 25 January 2005; Kilner, op. cit. note 4.

10. "Phillipines prepared to Launch Ethanol and Biodiesel Programs," ethanolmarketplace.com, 10 January 2005.

11. Philippines News Agency, "Govt Boosts Coco-diesel Production," *Manila Times*, 23 April 2004; Philippines News Agency, "RP, Thailand Agree on Reg'l Biofuels Standard," *Manila Times*, 1 September 2004.

12. Fulton et al., op. cit. note 1, p.15.

13. Ibid., p. 68.

14. Kilner, op. cit. note 4.

15. Iogen Corporation. "First Vehicle Fleet to Use Cellulose Ethanol," press release (Ottawa, ON, Canada: 15 December 2004).

16. Fulton et al., op. cit. note 1, p. 168.

17. IMPCO Technologies, "Market Segments," fact sheet (Cerritos, CA: 2003).

CLIMATE CHANGE INDICATORS ON THE RISE (pages 40–41)

1. Tim Whorf, Scripps Institution of Oceanography, e-mail to author, 18 January 2005.

2. Increase since 1959 calculated by Worldwatch with data from C. D. Keeling and T. P. Whorf, "Atmospheric Carbon Dioxide Record from Mauna Loa," Carbon Dioxide Information Analysis Center (CDIAC), Scripps Institution of Oceanography, University of California, La Jolla, CA, and from Whorf, op. cit. note 1; 35 percent increase from Arctic Climate Impact Assessment (ACIA), *Impacts of a Warming Arctic* (New York: Cambridge University Press, 2004), p. 2.

3. Tim Whorf, Scripps Institution of Oceanography, discussion with author, 21 January 2005.

4. Goddard Institute for Space Studies (GISS), NASA, "Global Land-Ocean Temperature Index in .01 C, base period 1951–1980 (January-December), 2005," at www.giss.nasa.gov/data/update/gistemp/GLB.Ts+dSST.txt.

5. "World Meteorologists Rank 2004 Fourth Warmest Year on Record," *Environment News Service*, 16 December 2004.

6. Ibid.

7. GISS, op. cit. note 4.

8. Maggie Fox, "Global Warming Effects Faster Than Feared—Experts," *Reuters*, 25 October 2004.

9. Shaoni Bhuttacharya, "Global Warming 'Kills 160,000 a Year'," *New Scientist*, 1 October 2003. For growing evidence see, for example, ACIA, op. cit. note 2; Camille Parmesan and Hector Galbraith, *Observed Impacts of Global Climate Change in the U.S.* (Washington, DC: Pew Center on Global Climate Change, November 2004); and Krista E. M. Galley, ed., "Global Climate Change and Wildlife in North

America," Technical Review 04-2 (Bethesda, MD: The Wildlife Society, December 2004).

10. Parmesan and Galbraith, op. cit. note 9.

11. J. R. Pegg, "Global Warming Linked to Increasing Drought," *Environment News Service*, 11 January 2005.

12. ACIA, op. cit. note 2, p. 8.

13. Ibid., pp. 10–11, 13.

14. Estimated by Worldwatch with data from BP, *Statistical Review of World Energy 2004* (London: 2004), from International Energy Agency, *Oil Market Report*, 18 January 2005, from "China Shelves 23 Power Stations, Citing Environment," *Reuters*, 20 January 2005, from Nao Nakanishi, "China Fuels Surge in Coal Demand, Prices," *Reuters*, 10 December 2004, from Energy Information Administration, *Quarterly Coal Report: July-September 2004* (Washington, DC: U.S. Department of Energy, 20 December 2004), and from G. Marland et al., "Global, Regional, and National Fossil Fuel CO_2 Emissions," in CDIAC, Oak Ridge National Laboratory, U.S. Department of Energy, *Trends: A Compendium of Data on Global Change* (Oak Ridge, TN: 2004).

15. Atmospheric concentrations from Whorf, op. cit. note 3; temperatures from J. T. Houghton et al., eds., *Climate Change 2001: The Scientific Basis*, Contribution of Working Group I to the Third Assessment Report of the Intergovernmental Panel on Climate Change (Cambridge, UK: Cambridge University Press, 2001).

16. Calculated by Worldwatch with data from BP, op. cit. note 14, and from Marland et al., op. cit. note 14.

17. Lila Buckley, "Carbon Emissions Reach Record High," *Eco-Economy Indicators* (Washington, DC: Earth Policy Institute, 2004).

18. "U.S. Greenhouse Gas Emissions Continue to Grow," *Environment News Service*, 15 December 2004.

19. Buckley, op. cit. note 17.

20. Ibid.

21. Daniel Wallis, "Russia Ratifies Kyoto, Starts on Feb. 16," *Reuters*, 19 November 2004.

22. *International Action Programme*, Official Outcome of the Intergovernmental Conference for Renewable Energies, Bonn, Germany, 30 August 2004, pp. 42-43.

23. Stuart Penson, "EU Launches Pioneering Emissions Trading Scheme," *Reuters*, 4 January 2005.

24. Ibid.

GLOBAL ECONOMY CONTINUES TO GROW (pages 44–45)

1. Angus Maddison, *The World Economy: Historical Statistics* (Paris: Organisation for Economic Co-operation and Development (OECD), 2003), p. 233, with updates from International Monetary Fund (IMF), *World Economic Outlook Database* (Washington, DC: September 2004). Note that this is a preliminary estimate from September 2004 and is subject to change.

2. IMF, op. cit. note 1.

3. Population data from U.S. Bureau of the Census, *International Data Base*, electronic database, Suitland, MD, updated 30 September 2004; gross world product from IMF, op. cit. note 1.

4. IMF, *World Economic Outlook 2004* (Washington, DC: 2004), pp. 3, 23–24.

5. Ibid., pp. 3, 30.

6. Ibid., pp. 3, 27.

7. Ibid.

8. Ibid., pp. 3, 32.

9. Ibid.

10. Ibid., pp. 3, 47.

11. Ibid., pp. 3, 53.

12. Ibid.

13. OECD, *The Well-being of Nations: The Role of Human and Social Capital* (Paris: 2001), pp. 10–11.

14. Ibid.

15. World Wide Fund for Nature (WWF), UNEP World Conservation Monitoring Centre, and Global Footprint Network, *Living Planet Report 2004* (Gland, Switzerland: WWF, 2004).

16. Ibid. The productivity of "global hectares" is an average based on the productivity of ecosystems used by humans.

17. Ibid.

18. Jason Venetoulis and Cliff Cobb, *The Genuine Progress Indicator 1950–2002 (2004 Update)* (Oakland, CA: Redefining Progress, 2004).

19. Ibid.

20. "China Plans to Set Up Green GDP System in 3–5 Years," *China Daily*, 12 March 2004.

21. "Blind Pursuit of GDP To Be Abandoned," *China Daily*, 5 March 2004.

WORLD TRADE RISES SHARPLY (pages 46–47)

1. Trade in goods and services from International Monetary Fund (IMF), *World Economic Outlook 2004*

(Washington, DC: September 2004), p. 28; trade in goods from IMF, *International Financial Statistics*, online database, January 2005.

2. Worldwatch calculation from IMF, *World Economic Outlook*, op. cit. note 1.

3. Ibid.

4. IMF, op. cit. note 1; gross world product from Angus Maddison, *The World Economy: Historical Statistics* (Paris: Organisation for Economic Co-operation and Development, 2003), p. 233, with updates from IMF, *World Economic Outlook Database* (Washington, DC: September 2004).

5. World Trade Organization (WTO), "2004 Trade Growth to Exceed 2003 Despite Higher Oil Price," press release (Geneva: 25 October 2004).

6. IMF, *World Economic Outlook*, op. cit. note 1; 2004 oil price from Centre for Energy, "Tetratechnologies, Inc. Announces 2005 Earnings Guidance and 2004 Fourth Quarter Earnings Estimate," The Woodlands, TX, 13 January 2005; price in 2003 from BP, *Statistical Review of World Energy 2004* (London: 2004), p. 14.

7. International Energy Agency, *Analysis of the Impact of High Oil Prices on the Global Economy* (Paris: Organisation for Economic Co-operation and Development, May 2004), p. 2.

8. World Bank, *Global Economic Prospects 2005* (Washington, DC: 2005), p. 8.

9. Ibid., p. 9.

10. Ibid.

11. World Bank, "Commodity Market Brief: Soybeans," online feature, at www.worldbank.org/global outlook.

12. Ibid.

13. U.N. Food and Agriculture Organization (FAO), "Agriculture Commodity Prices Continue Long-term Decline," press release (Rome: 15 February 2005).

14. WTO, "Day 5: Conference Ends Without Consensus," news brief (Cancun, Mexico: 14 September 2003).

15. International Centre for Trade and Sustainable Development, "Bridges—Cancun Special," No. 7 (Geneva: September-October 2003).

16. International Centre for Trade and Sustainable Development, "Overview of the July Package," Doha Briefing Series, Vol. 3 (Geneva: December 2004).

17. Ibid.

18. FAO, op. cit. note 13.

FOREIGN DIRECT INVESTMENT INFLOWS DECLINE (pages 48–49)

1. United Nations Conference on Trade and Development (UNCTAD), "Foreign Direct Investment (FDI)," at www.unctad.org/Templates/Page.asp?intItemID=3146&lang=1, viewed 05 January 2005; UNCTAD, *World Investment Report 2004: The Shift Towards Services* (Geneva: United Nations, 2004), p. 5.

2. UNCTAD, *FDI Database*, electronic database, updated September 2004.

3. Ibid. FDI inflows are the amount received by recipient (or host) countries while outflows are the amount invested by investor (or home) countries.

4. UNCTAD, "World FDI Flows Grew an Estimated 6% in 2004, Ending Downturn," press release (Geneva: 11 January 2005).

5. UNCTAD, op. cit. note 2.

6. Ibid. Low- and middle-income countries include all countries but those in Western Europe, the United States, Canada, Japan, Israel, Australia, and New Zealand.

7. UNCTAD, op. cit. note 4.

8. UNCTAD, *World Investment Report 2004*, op. cit. note 1, p. 367.

9. Ibid. Luxembourg received the most FDI of industrial countries, but most of this is actually "transshipped" elsewhere (see UNCTAD, *World Investment Report 2004*, op. cit. note 1, pp. xviii, 13).

10. UNCTAD, *World Investment Report 2004*, op. cit. note 1, p. 370.

11. Ibid., pp. 13–14.

12. Ibid., pp. 369–70. Hong Kong and Singapore, like Luxembourg, transfer a significant portion of FDI inflows to other countries (see UNCTAD, *World Investment Report 2004*, op. cit. note 1, p. 18).

13. UNCTAD, *World Investment Report 2004*, op. cit. note 1, pp. 367–71.

14. Ibid., p. 367.

15. Ibid., p. 372.

16. Ibid., pp. 372, 375.

17. Ibid., pp. 4–5.

18. Organisation for Economic Co-operation and Development (OECD), "Net ODA from DAC Countries from 1950 to 2003," *OECD Database*, updated December 2004; Americo Beviglia Zampetti and Torbjörn Fredriksson, "The Development Dimension of Investment Negotiations in the WTO," *The Journal of World Investment*, June 2003, pp. 399–450.

19. Official development assistance from OECD, op. cit.

note 18; FDI from UNCTAD, op. cit. note 2.

20. UNCTAD, *World Investment Report 2004*, op. cit. note 1, pp. 5–6.

21. UNCTAD, op. cit. note 2.

22. UNCTAD, *World Investment Report 2004*, op. cit. note 1, p. 30.

23. Ibid., p. 318.

24. Ibid.

25. Rosalie Gardiner, "Foreign Direct Investment: A Lead Driver for Sustainable Development?" *Towards Earth Summit 2002* (London: UNED Forum-UK Committee, 2001), p. 3.

26. UNCTAD, *World Investment Report 2004*, op. cit. note 1, p. xxiii.

27. Ibid.; Gardiner, op. cit. note 25, p. 3.

28. Gardiner, op. cit. note 25, p. 3.

29. Zampetti and Fredriksson, op. cit. note 18.

30. UNCTAD, *World Investment Report 2004*, op. cit. note 1, p. xxiii.

WEATHER-RELATED DISASTERS NEAR A RECORD (pages 50–51)

1. Angelika Wirtz, Munich Reinsurance Company (Munich Re), e-mail to Janet Sawin, 15 February 2005.

2. Angelika Wirtz, Munich Re, e-mail to Janet Sawin, 28 January 2003; Wirtz, op. cit. note 1.

3. Wirtz, op. cit. note 1; Wirtz, op. cit. note 2.

4. Wirtz, op. cit. note 2; Wirtz, op. cit. note 1.

5. AFP, "Missing Expected to Take Tsunami Toll past 280,000," *ABC News Online*, 25 January 2005.

6. Munich Re, "The Ten Largest Natural Catastrophes in 2004," at www.munichre.com, viewed 20 January 2005.

7. Concern Worldwide, "Nearly 75% of Bangladesh Underwater; Concern Warning of Increasing Humanitarian Crisis," press release (New York: 27 July 2004).

8. Munich Re, op. cit. note 6.

9. Lester R. Brown, *Eco-Economy* (New York: W.W. Norton & Company, 2001), p. 169; Doug Rekenthaler, "China Floods Exacerbated by Man's Impact on Land, Climate," *Disaster Relief*, 8 October 2004.

10. Dan Bjarnason, "Deforestation in Haiti," *CBC News Online*, 1 October 2004.

11. Wirtz, op. cit. note 2.

12. U.N. Environment Programme, *Environmental Emergencies Newsletter* (Nairobi), September 2004.

13. Janet Sawin, "Severe Weather Events on the Rise," in Worldwatch Institute, *Vital Signs 2003* (New York: W.W. Norton & Company, 2005), p. 93.

14. Munich Re, op. cit. note 6.

15. Wirtz, op. cit. note 1.

16. Munich Re, op. cit. note 6.

17. Wirtz, op. cit. note 2; Wirtz, op. cit. note 1.

18. Wirtz, op. cit. note 1; Wirtz, op. cit. note 2.

19. Wirtz, op. cit. note 2; Munich Re, op. cit. note 6.

20. According to Essam El-Hinnawi of the Natural Resources and Environmental Institute in Cairo, cited in Rhoda Margesson, "Environmental Refugees," in Worldwatch Institute, *State of the World 2005* (New York: W.W. Norton & Company, 2005), p. 40.

21. Töpfer cited in Mark Townsend, "Environmental Refugees," *The Ecologist*, June 2002.

22. Bjarnason, op. cit. note 10.

STEEL SURGING (pages 52–53)

1. International Iron and Steel Institute (IISI), "The Largest Steel Producing Countries, 2004," at www.worldsteel.org, viewed 2 February 2005.

2. IISI, *Steel Statistical Yearbook 2004* (Brussels: 2005), pp. 10–11.

3. Ibid.

4. IISI, "World Produces 1.05 Billion Tonnes of Steel in 2004," press release (Brussels: 19 January 2005).

5. IISI, op. cit. note 2, p. 12.

6. IISI, op. cit. note 1.

7. Asia's share from IISI, op. cit. note 2, p. 12; 48 percent calculated from IISI, op. cit. note 4.

8. IISI, op. cit. note 2, p. 83.

9. Global Insight, Inc., "World Outlook for Key Drivers of Demand for Iron and Steel," fact sheet, at www.globalinsight.com, viewed 5 February 2005.

10. Ibid.

11. Michael Fenton, "Iron and Steel Scrap," *Mineral Commodity Summaries* (Washington, DC: U.S. Geological Survey, January 2005), pp. 88–89.

12. IISI, op. cit. note 2, pp. 39–41.

13. Fenton, op. cit. note 11, pp. 88–89.

14. Ibid.

15. Ibid.

16. "International Iron and Steel Institute Shares Short-Range Outlook," *American Recycler*, June 2004.

17. Ibid.

18. Ian Porter, "Nissan to Put Japanese Production on Hold," *The Age*, 29 November 2004.

19. Christopher Davis, "Rise in Global Steel Demand

Notes

May Impact Coke Shortage," *Pittsburgh Business Times*, 26 January 2004.

20. John Anton, "Steel Overview," Global Insight, Inc., undated.

21. Ibid.

22. John Anton, "China's Dominant Role in the Global Steel Market," Global Insight, Inc., undated.

23. Elaine Kurtenbach, "China's Buying Spree Spreads Global Impact," *Detroit News*, 22 October 2004.

24. Ibid.

VEHICLE PRODUCTION SETS NEW RECORD (pages 56–57)

1. Colin Couchman, Global Insight Automotive Group, e-mail to author, 13 January 2005.

2. Ibid.; Global Insight Automotive Group, *Global Sales of Light Vehicles by Region & Country December 2003* (London: 2003), and earlier editions (as DRI-WEFA); American Automobile Manufacturers Association, *World Motor Vehicle Facts and Figures 1998* (Washington, DC: 1998).

3. Couchman, op. cit. note 1.

4. Ibid.

5. Calculated from Ward's Communications, *Ward's Motor Vehicle Facts & Figures 2004* (Southfield, MI: 2004), pp. 14, 47–49. Ward's includes Mexico in its definition of North America and Turkey as part of Western Europe. For the purposes of this calculation, these two countries were not included.

6. Ward's Communications, op. cit. note 5.

7. Oak Ridge National Laboratory (ORNL), *Transportation Energy Data Book 24* (Oak Ridge, TN: 2004), Figure 3.1.

8. Ibid.

9. Ibid., Tables 11.1 and 11.5.

10. Lew Fulton, International Energy Agency, e-mail to author, 20 December 2001.

11. Danny Hakim, "At $2 a Gallon, Gas Is Still Worth Guzzling," *New York Times*, 16 May 2004.

12. Danny Hakim, "Average U.S. Car Is Tipping Scales at 4,000 Pounds," *New York Times*, 5 May 2004.

13. ORNL, op. cit. note 7, Table 4.15.

14. Ward's Communications, op. cit. note 5, p. 56.

15. "Annual Vehicle Distance Traveled in Miles and Related Data, 1936–1995," in U.S. Federal Highway Administration (FHWA), *Highway Statistics Summary to 1995* (Washington, DC: 1997).

16. "Annual Vehicle Distance Traveled in Miles and Related Data, 2003," in FHWA, *Highway Statistics*

2003 (Washington, DC: 2004).

17. Ibid.

18. The distance from Earth to the Sun is 149.6 million kilometers; *Atlas of the Skies* (Cobham, Surrey, U.K.: TAJ Books, 2003), p. 139.

19. ORNL, op. cit. note 7, Table 2.6.

20. Ibid., Table 1.12 and Figure 1.7.

21. "Driving the Economy," *National Geographic*, February 2005.

22. Keith Bradsher, "China Sets Its First Fuel-Economy Rules," *New York Times*, 23 September 2004.

23. Ibid.

24. "Toyota's Hybrid Car Sales Total 280,000," *Japan Today*, 19 January 2005.

25. Danny Hakim, "Automakers Unveil Plans for More Hybrid Models," *New York Times*, 6 January 2004.

26. ORNL, op. cit. note 7, Table 6.5.

27. "US Prepares for Hybrid Onslaught," *BBC News Online*, 11 January 2005.

28. Hakim, op. cit. note 25; Bradley Berman, "Hybrid Cars: The Slow Drive to Energy Security," *Energy Security* (Institute for the Analysis of Global Security), 13 January 2005.

29. Dee-Ann Durbin, "Hybrid Car Sales Expected to Slow," *Associated Press*, 11 January 2005.

30. ORNL, op. cit. note 7, Table 6.1.

BICYCLE PRODUCTION RECOVERS (pages 58–59)

1. United Nations, *The Growth of World Industry,* 1969 Edition, Vol. II (New York: 1971); United Nations, *Yearbooks of Industrial Statistics,* 1979 and 1989 Editions, Vol. II (New York: 1981 and 1991); *Interbike Directory*, various years; United Nations, *Industrial Commodity Statistics Yearbook* (New York: various years); *Bicycle Retailer and Industry News Directory*, various years.

2. *Interbike Directory*, op. cit. note 1; United Nations, *Industrial Commodity Statistics Yearbook*, op. cit. note 1; *Bicycle Retailer and Industry News Directory*, op. cit. note 1.

3. *Bicycle Retailer and Industry News Directory*, op. cit. note 1.

4. Ibid.

5. Ibid.

6. Ibid.

7. *Interbike Directory*, op. cit. note 1; United Nations, *Industrial Commodity Statistics Yearbook*, op. cit. note 1; *Bicycle Retailer and Industry News Directory* , op. cit. note 1.

8. "Stop World Bank Pressure to Ban Rickshaws," *Sustainable Transport*, winter 2004, p. 6.

9. John Pucher and John L. Renne, "Socioeconomics of Urban Travel: Evidence from the 2001 NHTS," *Transportation Quarterly*, summer 2003, p. 50.

10. Aimée Gauthier, "Using Bicycles to Save Lives," *Sustainable Transport*, winter 2004, pp. 8–11.

11. Ibid.

12. Carlos Felipe Pardo, "Feeding Bogotá's TransMilenio with Non-Motorized Transport," *Sustainable Transport*, winter 2004, pp. 6–7.

13. Ibid.

14. John Pucher and Lewis Dijkstra, "Promoting Safe Walking and Cycling to Improve Public Health: Lessons from the Netherlands and Germany," *American Journal of Public Health*, September 2003, pp. 1509–16.

15. Pucher and Renne, op. cit. note 9, p. 50.

16. Pucher and Dijkstra, op. cit. note 14, pp. 1509–16.

17. Ibid.

18. Ibid.

19. Ibid.

20. Ibid.

AIR TRAVEL SLOWLY RECOVERING (pages 60–61)

1. International Civil Aviation Organization (ICAO), "World Air Passenger Traffic to Rebound Strongly in 2004 and Continue to Grow in 2005 and 2006," press release (Montreal: 22 September 2004); increase in air travel based on passenger-kilometers performed.

2. ICAO Statistics Section, e-mail to author, 20 January 2005. Figures for 1950–69 do not include states formerly within the Soviet Union; figures for 2003 are provisional ICAO estimates.

3. ICAO Statistics Section, op. cit. note 2.

4. ICAO Statistics Section, e-mail to author, 3 February 2005; John Whitelegg and Howard Cambridge, *Aviation and Sustainability* (Stockholm: Stockholm Environment Institute, 2004), p. 7.

5. ICAO Statistics Section, op. cit. note 2; ICAO, op. cit. note 1.

6. Rebecca Bowe et al., "Flying the Dirty Skies," *E Magazine*, September/October 2004; North America represented 34 percent of the ton-kilometers performed and 38 percent of the passengers carried in 2003, according to ICAO Statistics Section, op. cit. note 4.

7. ICAO Statistics Section, op. cit. note 4.

8. .Whitelegg and Cambridge, op. cit. note 4, p. 7.

9. Ibid.

10. Boeing, *Current Market Outlook 2004*, at www.boeing.com/commercial/cmo.

11. Ibid.

12. Ibid.

13. Organisation for Economic Co-operation and Development, *Policy Instruments for Achieving Environmentally Sustainable Transport* (Paris: 2002).

14. "Oil Prices to Hit Airline Profits," *BBC News*, 7 June 2004; "High Oil Costs Hit More Airlines," *BBC News*, 18 August 2004.

15. Intergovernmental Panel on Climate Change (IPCC), *Aviation and the Global Atmosphere: Summary for Policy Makers* (Cambridge, U.K.: Cambridge University Press, 1999); Whitelegg and Cambridge, op. cit. note 4, p. 7.

16. IPCC, op. cit. note 15.

17. Ibid. These estimates do not take into account possible changes in cirrus clouds.

18. IPCC, op. cit. note 15; Bowe et al., op. cit. note 6; Kyoto Protocol to the United Nations Framework Convention on Climate Change, Article 2.2.

19. D. J. Travis et al., "Contrails Reduce Daily Temperature Range," *Nature*, 8 August 2002, p. 601.

20. Whitelegg and Cambridge, op. cit. note 4, pp. 17–18; ICAO Statistics Section, op. cit. note 4.

21. Whitelegg and Cambridge, op. cit. note 4, pp. 17–18.

22. Ibid.; ICAO Statistics Section, op. cit. note 4.

23. Whitelegg and Cambridge, op. cit. note 4, p. 8.

24. Bowe et al., op. cit. note 6.

25. Whitelegg and Cambridge, op. cit. note 4, p. 8.

POPULATION CONTINUES ITS STEADY RISE (pages 64–65)

1. U.S. Bureau of the Census, *International Data Base*, electronic database, Suitland, MD, updated 30 September 2004.

2. Ibid.

3. Ibid.

4. U.N. Population Division, *World Population Prospects: The 2002 Revision* (New York: 2003).

5. Ibid.

6. Ibid.

7. Ibid.

8. Ibid.

9. Gary Gardner, Eric Assadourian, and Radhika Sarin, "The State of Consumption Today," in Worldwatch

Institute, *State of the World 2004* (New York: W.W. Norton & Company, 2004).

10. Ibid., p. 5.

11. Ibid.

12. U.N. Population Division, *World Urbanization Prospects: The 2003 Revision* (New York: 2004).

13. Ibid.

14. Ibid.

15. U.N. Population Division, op. cit. note 4.

16. Ibid.

17. Ibid.

18. U.N. Population Fund (UNFPA), *State of World Population 2004* (New York: 2004).

19. Ibid.

20. Ibid.

21. World Health Organization (WHO), UNICEF, and UNFPA, *Maternal Mortality in 2000: Estimates Developed by WHO, UNICEF, and UNFPA* (Geneva: WHO, 2003).

NUMBER OF REFUGEES DECLINES (pages 66–67)

1. U.N. High Commissioner for Refugees (UNHCR), *Refugees by Numbers 2004* (Geneva: 2004) and earlier editions.

2. United Nations Relief and Works Agency for Palestine Refugees in the Near East, "UNRWA in Figures" and other documents available at www.unrwa.org.

3. UNHCR, op. cit. note 1, p. 14.

4. Ibid.

5. "Hundreds of Thousands of Afghan Refugees Likely to Return Home this Year—UN," *UN News Service*, 17 January 2005.

6. All numbers, except for Myanmar, from UNHCR, *2003 Global Refugee Trends* (Geneva: 2004), p. 3; Myanmar from U.S. Committee for Refugees, *World Refugee Survey 2004* (Washington, DC: 2004), Table 6.

7. UNHCR, op. cit. note 6, p. 3.

8. Ibid.

9. UNHCR, op. cit. note 1, pp. 6, 13.

10. Ibid., p. 5.

11. The Global IDP Project, *Internal Displacement: A Global Overview of Trends and Developments in 2003* (Geneva: February 2004), p. 4.

12. UNHCR, op. cit. note 1, p. 6.

13. International Federation of Red Cross and Red Crescent Societies, *World Disaster Report 2004* (Bloomfield, CT: Kumarian Press, 2004), Table 16; time series from ibid. and from U.S. Committee for Refugees,

op. cit. note 6, and earlier editions.

14. Ibid.

15. Ibid.

16. The Global IDP Project, op. cit. note 11, p. 4.

17. Rhoda Margesson, "Environmental Refugees," in Worldwatch Institute, *State of the World 2005* (New York: W.W. Norton & Company, 2005), p. 40.

18. "Creeping Desertification: The Cause and Consequence of Poverty," *Environment News Service*, 18 June 2004.

19. The figure of 70 million is the sum of the international refugees (including Palestinian refugee communities), asylum seekers, internally displaced people, and El-Hinnawi's estimate for environmental refugees.

HIV/AIDS CRISIS WORSENING WORLDWIDE (pages 68–69)

1. Estimates based on Joint United Nations Programme on HIV/AIDS (UNAIDS), *AIDS Epidemic Update* (Geneva: various years).

2. UNAIDS, *AIDS Epidemic Update: December 2004* (Geneva: December 2004), p. 1; UNAIDS, op. cit. note 1.

3. UNAIDS, *2004 Report on the Global AIDS Epidemic* (Geneva: 2004).

4. Richard P. Cincotta, Robert Engelman, and Daniele Anastasion, *The Security Demographic: Population and Civil Conflict After the Cold War* (Washington, DC: Population Action International (PAI), 2003).

5. Ibid.

6. Ibid.

7. Estimate of 53 includes 49 countries with HIV prevalence at 1 percent or greater, plus Russia, United States, India, and China, per U.N. Population Division, *World Population Prospects: The 2002 Revision (Highlights)* (New York: United Nations, 2003); U.N. Development Programme, "HIV/AIDS Crisis Drives Down Life Expectancy, Human Development Rankings in Sub-Saharan Africa," press release (Bangkok: 14 July 2004).

8. U.N. Population Division, op. cit. note 7.

9. UNAIDS, *Epidemiological Fact Sheets on HIV/AIDS and Sexually Transmitted Infections* (Geneva: 2002); R. Greener, "AIDS and Macroeconomic Impact," in S. Fortsythe, ed., *State of the Art: AIDS and Economics* (Washington, DC: International AIDS-Economics Network, 2002), pp. 49–54.

10. UNAIDS, op. cit. note 1.

11. Laure Belot, "Le Sida, Un Risque Croissant Pour Les Entreprises En Afrique," *Le Monde*, 22 May 2004.

12. Estimate of 15 percent is from PAI and is based on the following: T. Butcher, "HIV and Lack of Funds Paralyse S. Africa's Army," at news.telegraph.co.uk, 16 July 2002; L. Heinecken, "Strategic Implications of HIV/AIDS in South Africa," *Conflict, Security and Development*, vol. 1, no. 1 (2001), pp. 109–15; UNAIDS, *AIDS and the Military: UNAIDS Point of View* (Geneva: May 1998); International Crisis Group, *HIV/AIDS as a Security Issue* (Washington, DC: June 2001); "AIDS: An Intelligence Issue," *The Namibian*, 13 February 2001; and Armed Forces Medical Intelligence Center, *Impact of HIV/AIDS on Military Forces: Sub-Saharan Africa* (Washington, DC: Defense Intelligence Agency, 2000).

13. Exceeding civilian rates from Ugboga Adaji Nwokoji and Ademola J Ajuwon, "Knowledge of AIDS and HIV Risk-Related Sexual Behavior Among Nigerian Naval Personnel," *BMC Public Health*, 21 June 2004; Poverty Reduction Forum, *Zimbabwe Human Development Report 2003* (Mt. Pleasant, Zimbabwe: Institute of Development Studies, University of Zimbabwe, 2004), p. 17.

14. International Labour Organization (ILO), *HIV/AIDS and Work: Global Estimates, Impact and Response* (Geneva: July 2004).

15. Cincotta, Engelman, and Anastasion, op. cit. note 4.

16. ILO, op. cit. note 14, p. 13.

17. UNAIDS, op. cit. note 2.

18. By definition, an AIDS orphan is under 15 years of age and has lost his or her mother or both parents from an AIDS-related cause, per UNAIDS, *Children and Young People in a World of AIDS* (Geneva: 2001); UNICEF, UNAIDS, and USAID, *Children on the Brink 2004: A Joint Report of New Orphan Estimates and a Framework for Action* (New York: July 2004).

19. U.N. Population Division, op. cit. note 7.

20. Barbara Crossette, "AIDS Catastrophe Looms in India," *U.N. Wire*, 14 July 2004; UNAIDS, op. cit. note 3.

21. UNAIDS, op. cit. note 2, p. 2.

22. Peter Baken, "Russia Sees an AIDS 'Explosion'," *Washington Post*, 13 June 2004.

23. C. Ruhl et al., *Computer-based Model: The Economic Consequences of HIV in Russia* (Moscow: World Bank Group in Russia, June 2002); Cincotta, Engelman, and Anastasion, op. cit. note 4.

24. UNAIDS, op. cit. note 2, p. 5.

25. Ibid.

CIGARETTE PRODUCTION DROPS (pages 70–71)

1. U.S. Department of Agriculture (USDA), *Production, Supply, and Distribution*, electronic database, updated 30 September 2004.

2. Ibid.; population data from U.S. Bureau of the Census, *International Data Base*, electronic database, Suitland, MD, updated 30 September 2004.

3. USDA, op. cit. note 1; Census Bureau, op. cit. note 2.

4. USDA, op. cit. note 1.

5. Ibid.

6. Ibid.

7. Ibid.

8. Ibid. Consumption of cigarettes is a residual number based on total production plus imports minus exports. Thus, this number includes stockpiled cigarettes and cannot factor in discrepancies due to smuggling.

9. USDA, op. cit. note 1.

10. Ibid.

11. Ibid.; Census Bureau, op. cit. note 2. Japan imported 83 billion cigarettes and exported 20 billion in 2004.

12. Figure of 30 percent from Masaoki Nagahama, "Japan: Tobacco and Products Annual 2004," *Global Agriculture Information Network Report* (Washington, DC: USDA, Foreign Agricultural Service, 1 May 2004).

13. Majid Ezzati and Alan Lopez, "Estimates of Global Mortality Attributable to Smoking in 2000," *The Lancet*, 13 September 2003, pp. 847–52.

14. Population from Judith Mackay and Michael Eriksen, *The Tobacco Atlas* (Geneva: World Health Organization (WHO), 2002), p. 36; aggressive marketing from Pan American Health Organization, *Profits Over People: Tobacco Industry Activities to Market Cigarettes and Undermine Public Health in Latin America and the Caribbean* (Washington, DC: November 2002).

15. Ezzati and Lopez, op. cit. note 13.

16. Ibid.

17. Deaths in 2030 from Richard Peto and Alan D. Lopez, "Future Worldwide Health Effects of Current Smoking Patterns," in C. Everett Koop, Clarence E. Pearson, and M. Roy Schwarz, eds., *Critical Issues in Global Health* (San Francisco, CA: Jossey-Bass, 2001), pp. 154–61; 7 in 10 from C. K. Gajalakshmi et al., "Global Patterns of Smoking and Smoking-attributable Mortality," in Prabhat Jha and Frank

Chaloupka, eds., *Tobacco Control in Developing Countries* (Oxford: Oxford University Press, 2000), p. 35.

18. "All Eyes on Ireland's Smoking Ban," *BBC News*, 29 March 2004.

19. "Scotland Smoking Ban To Go Ahead," *BBC News*, 11 November 2004.

20. Ibid.; "Norwegians Ban Smoking in Bars," *BBC News*, 1 June 2004; Elisabetta Povoledo, "Italy's Ban On Smoking Gets Off To a Fuming Start," *International Herald Tribune*, 11 January 2005.

21. David P. Hopkins et al., "Reviews of Evidence Regarding Interventions to Reduce Tobacco Use and Exposure to Environmental Tobacco Smoke," *American Journal of Preventive Medicine*, vol. 20, issue 2S (2001), pp. 16–66.

22. "Living in an Anti-Smoking Climate," *BBC News*, 10 November 2004.

23. "Curbing Tobacco Use in Poland," in Ruth Levine et al., eds., *Millions Saved: Proven Successes in Global Health* (Washington, DC: Center for Global Development, 2004), pp. 113–20.

24. Ibid.

25. Ibid.

26. WHO, "WHO Tobacco Treaty Set to Become Law, Making Global Public Health History," press release (Geneva: 1 December 2004).

27. United Nations, *WHO Framework Convention on Tobacco Control*, adopted by Fifty-sixth World Health Assembly, Geneva, 21 May 2003.

VIOLENT CONFLICTS UNCHANGED (pages 74–75)

1. Arbeitsgemeinschaft Kriegsursachenforschung (AKUF), "Die 'Normalität' des Kriegsgeschehens. Zahl der Kriege Konstant Geblieben," press release (Hamburg, Germany: University of Hamburg, 13 December 2004); Wolfgang Schreiber, AKUF, e-mail to author, 22 December 2004.

2. AKUF, op. cit. note 1; Schreiber, op. cit. note 1.

3. AKUF, op. cit. note 1; AKUF, "Das Kriegsgeschehen 2003 im Überblick," at www.sozialwiss.uni-hamburg.de/publish/Ipw/Akuf/kriege_aktuell.htm, viewed 21 December 2004; Schreiber, op. cit. note 1.

4. AKUF, op. cit. note 1.

5. Ibid.

6. Ibid.

7. Taylor B. Seybolt, "Measuring Violence: An Introduction to Conflict Data Sets," in Stockholm International Peace Research Institute, *SIPRI Yearbook 2002: Armaments, Disarmament and Inter-*

national Security (New York: Oxford University Press, 2002), pp. 84–85.

8. Nils Petter Gleditsch et al., "Armed Conflict: 1946–2001: A New Dataset," *Journal of Peace Research*, vol. 39, no. 5 (2002), pp. 615–37, with updates for 2002 and 2003 from International Peace Research Institute, Oslo, Web site, at www.prio.no/cwp/armedconflict/current/conflict_list_1946–2003.pdf. In Figure 2, armed conflicts are the sum of "minor armed conflicts" (at least 25 battle-related deaths in a year), "intermediate conflicts" (more than 1,000 battle-related deaths during the course of the conflict), and "wars" (at least 1,000 battle-related deaths in a given year). For most of the "unclear cases," there is uncertainty about whether fatalities during the reporting year surpassed 25 (in addition to other outstanding definitional questions).

9. Heidelberger Institut für Internationale Konfliktforschung (HIIK), *Konfliktbarometer 2004* (Heidelberg, Germany: Institute for Political Science, University of Heidelberg, 2004), p. 3.

10. Ibid.

11. Ibid., p. 5.

12. Ibid., p. 7.

13. Ibid.

14. Ibid., p. 8.

15. Ibid.

16. Bethany Lacina and Nils Petter Gleditsch, *Monitoring Trends in Global Combat: A New Dataset of Battle Deaths* (Oslo: Centre for the Study of Civil War, International Peace Research Institute, Oslo, 2004).

17. International Rescue Committee and Burnet Institute, *Mortality in the Democratic Republic of Congo: Results from a Nationwide Survey* (New York and Melbourne: December 2004), p. iii.

18. Lacina and Gleditsch, op. cit. note 16, p. 23.

19. Ibid., p. 32.

20. David Nabarro, "Sudan: Mortality Projections for Darfur," media briefing notes (Geneva: World Health Organization, 15 October 2004).

21. Les Roberts et al., "Mortality Before and After the 2003 Invasion of Iraq: Cluster Sample Survey," *The Lancet*, 6 November 2004, p. 1857.

22. Ibid.

MILITARY EXPENDITURES SURGE (pages 76–77)

1. Unless otherwise noted, all monetary terms are expressed in 2003 dollars. Elisabeth Sköns et al.,

"Military Expenditure," in Stockholm International Peace Research Institute (SIPRI), *SIPRI Yearbook 2004: Armaments, Disarmament and International Security* (New York: Oxford University Press, 2004), Appendix 10A, p. 340; Rita Tulberg, "World Military Expenditure," *Bulletin of Peace Proposals*, no. 3–4, 1986; Bonn International Center for Conversion, *Conversion Survey* (Baden-Baden, Germany: Nomos Verlagsgesellschaft, annual), various editions; SIPRI, *SIPRI Yearbook* (New York: Oxford University Press, annual), various editions. The time series presented in Figure 1 has been compiled from a variety of sources, and there may be some incompatibilities in terms of underlying methodologies. The intent is to provide a rough indication of longer-term trends.

2. Calculated by author.

3. Calculated by author on basis of data in Sköns et al., op. cit. note 1.

4. Ibid.

5. Ibid., p. 312.

6. "Fiscal Year 2005 Pentagon Budget Request," Center for Arms Control and Non-Proliferation, at 64.177.207.201/static/budget/annual/fy05/topline.cfm, viewed 5 January 2005. These figures include nuclear weapons–related budgets under the purview of the U.S. Department of Energy. Time series from U.S. Department of Defense, Office of the Undersecretary of Defense (Comptroller), *National Defense Budget Estimates*, fiscal year 2003 and 2004 editions (Washington, DC: March 2002 and March 2003), and from Sköns et al., op. cit. note 1, Tables 10.5 and 10.6.

7. Sköns et al., op. cit. note 1, p. 319.

8. Ibid.

9. Ibid., p. 315.

10. Bryan Bender, "War Funding Request May Hit $100 Billion," *Boston Globe*, 15 December 2004; Phyllis Bennis and the IPS Iraq Task Force, *A Failed "Transition": The Mounting Costs of the Iraq War* (Washington, DC: Institute for Policy Studies and Foreign Policy In Focus, 2004), p. 15.

11. Don Gonyea, "Bush to Request $80 Billion for Iraq, Afghanistan," *Morning Edition*, National Public Radio, 25 January 2005; Tom Shanker and Eric Schmitt, "Pentagon Budget Up: War Cost Is Excluded," *New York Times*, 8 February 2005.

12. Sköns et al., op. cit. note 1, p. 312.

13. Ibid.

14. Ibid., pp. 312–13.

15. Ibid., p. 305.

16. Ibid. The comparison is for 2001. Given soaring military spending since then, this gap has widened further.

17. Sköns et al., op. cit. note 1, p. 308.

18. Ibid.

19. These are principally United Nations–generated estimates, as collected by the World Game Institute, "Global Priorities," wall chart, 2000, available at www.worldgame.org, and by Bennis and the IPS Iraq Task Force, op. cit. note 10, p. 54.

20. Malaria spending from "War on Terror Drains Funds from Poverty Fight," *Agence France Presse*, 2 December 2003.

21. Hilary French, Gary Gardner, and Erik Assadourian, "Laying the Foundations for Peace," in Worldwatch Institute, *State of the World 2005* (New York: W.W. Norton & Company, 2005), p. 170.

22. Ibid.

PEACEKEEPING EXPENDITURES SOAR
(pages 78–79)

1. U.N. Department of Public Information (UNDPI), "United Nations Peacekeeping Operations. Background Note" (New York: 31 December 2004, and earlier editions); Worldwatch Institute database. All dollar amounts are in 2003 dollars.

2. U.N. Department of Peacekeeping Operations (UNDPKO), "Monthly Summary of Contributors," at www.un.org/Depts/dpko/dpko/contributors/index.htm, viewed 15 January 2005; personnel number also based on William Durch, Henry Stimson Center, Washington, DC, e-mail to author, 9 January 1996, and on Global Policy Forum, at www.globalpolicy.org/security/peacekpg/data/pkomctab.htm, viewed 22 December 2004.

3. UNDPI, op. cit. note 1.

4. Ibid.

5. Ibid.

6. Author's calculation, based on data from UNDPKO, op. cit. note 2.

7. Ibid.

8. Bates Gill, "China's New Security Multilateralism and Its Implications for the Asia-Pacific Region," in Stockholm International Peace Research Institute (SIPRI), *SIPRI Yearbook 2004* (New York: Oxford University Press, 2004), pp. 222–25.

9. Calculated from UNDPI, op. cit. note 1.

10. Ibid.

11. Ibid.
12. United Nations Security Council, Resolution 1565, 1 October 2004.
13. "More Than 10,000 Troops Proposed for UN Peace-Support Mission for Sudan," *UN News Service*, 3 February 2005.
14. "Status of Contributions to the Regular Budget, International Tribunals, Peacekeeping Operations and Capital Master Plan for the Biennium 2002–2003 as at 31 December 2004," from Mark Gilpin, Chief of United Nations Contributions Service, New York, letter to author, 17 January 2005; Global Policy Forum, "US vs. Total Debt to the UN: 2004," at www.globalpolicy.org/finance/tables/core/un-us-04.htm, and "US vs. Total Debt to the UN: 2003," at www.globalpolicy.org/finance/tables/core/un-us-03.htm, both viewed 21 December 2004.
15. "Status of Contributions," op. cit. note 14; Global Policy Forum, "Contributions Owing to the UN for Peacekeeping Operations: 2004," at www.global policy.org/finance/tables/pko/due2004.htm, viewed 21 December 2004.
16. "Status of Contributions," op. cit. note 14; Global Policy Forum, op. cit. note 15.
17. "Status of Contributions," op. cit. note 14.
18. "U.N. Welcomes Indonesia's Plan for ASEAN Peace-keeping Force," *UN Wire*, 26 February 2004.
19. "African Countries Agree to Standby Force to Halt Conflicts," *UN Wire*, 1 March 2004.
20. Fanen Chiahemen, "G-8 Countries to Train 50,000 African Peacekeepers," *UN Wire*, 9 June 2004.
21. Renata Dwan and Sharon Wiharta, "Multilateral Peace Missions," in SIPRI, op. cit. note 8, pp. 149–90; International Institute for Strategic Studies (IISS), "The 2004 Chart of Armed Conflict," wall chart distributed with IISS, *The Military Balance 2004–2005* (London: Oxford University Press, 2004); Worldwatch Institute database.
22. Worldwatch estimates, based primarily on data from Dwan and Wiharta, op. cit. note 21, and on IISS, "The 2004 Chart," op. cit. note 21.
23. Dwan and Wiharta, op. cit. note 21; IISS, "The 2004 Chart," op. cit. note 21.
24. Nicholas Wood, "Europeans Set to Succeed NATO in Bosnia," *New York Times*, 2 December 2004; Ian Black, "Peacekeeping Forces Power Agenda," (London) *The Guardian*, 2 December 2004.
25. "EU Launches Crisis Police Force," *BBC News Online*, 17 September 2004. The five countries are France, Italy, Spain, Portugal, and the Netherlands.
26. Calculated from IISS, *The Military Balance 2004–2005*, op. cit. note 21.
27. Ibid.

MIXED PROGRESS ON REDUCING NUCLEAR ARSENALS (pages 80–81)

1. Robert S. Norris and Hans M. Kristensen, "Global Nuclear Stockpiles, 1945–2002," *Bulletin of the Atomic Scientists*, November/December 2002, pp. 103–04; various "Nuclear Notebook" articles on the Web site of *The Bulletin of the Atomic Scientists*. The numbers here have been revised from those reported in earlier editions of *Vital Signs*. Warhead data are estimates because a great deal of information about these arsenals remains a government secret.
2. Hans M. Kristensen, "World Nuclear Forces," in Stockholm International Peace Research Institute (SIPRI), *SIPRI Yearbook 2004: Armaments, Disarmament and International Security* (New York: Oxford University Press, 2004), p. 633.
3. Ibid., p. 628.
4. Calculated from Norris and Kristensen, op. cit. note 1, and from "Nuclear Notebook," op. cit. note 1.
5. Kristensen, op. cit. note 2, p. 629.
6. Estimate of 100 from "Nuclear Numbers," www .carnegieendowment.org/npp/numbers/default.cfm, last updated April 2004; 200 warheads estimate from Kristensen, op. cit. note 2, pp. 645–46.
7. Kristensen, op. cit. note 2, p. 629.
8. Shannon Kile, "Nuclear Arms Control and Non-Proliferation," in SIPRI, op. cit. note 2, pp. 621, 623.
9. Ibid., p. 622.
10. Norris and Kristensen, op. cit. note 1, pp. 103–04.
11. Council on Foreign Relations, "Nuclear Weapons. Editorial Briefing with Robert Nelson," at www.cfr.org/pub6435/robert_w_nelson/nuclear_we aponseditorial_briefing_with_robert_nelson.php, viewed 30 December 2004.
12. Institute for Science and International Security, "Summary Table: Production and Status of Military Stocks of Fissile Material, End of 2003 (in Tonnes)," at www.isis-online.org/mapproject/supplements.html.
13. David Albright and Kimberly Kramer, "Fissile Material: Stockpiles Still Growing," *Bulletin of the Atomic Scientists*, November/December 2004, pp. 14–15.
14. Ibid.
15. Ibid.
16. Ibid.

17. Reaching Critical Will, "Fissile Materials Cut-Off Treaty," at www.reachingcriticalwill.org/legal/fmct.html, viewed 31 December 2004; Dafna Linzer, "U.S. Shifts Stance on Nuclear Treaty," *Washington Post*, 31 July 2004.

18. Barbara Crossette, "5 Nuclear Powers Agree on Stronger Pledge to Scrap Arsenals," *New York Times*, 22 May 2000; Kristensen, op. cit. note 2, p. 628.

19. Excerpts from the secret Nuclear Posture Review were leaked and are available at www.globalsecurity.org/wmd/library/policy/dod/npr.htm.

20. Kristensen, op. cit. note 2, p. 631; Kile, op. cit. note 8, p. 603.

21. Arms Control Association, "Arms Control Association Applauds Lawmakers' Move to Cut Funding for Costly and Counterproductive Nuclear Weapons Projects," press release (Washington, DC: 22 November 2004).

22. "Proliferation Security Initiative," at www.globalsecurity.org/military/ops/psi.htm, viewed 29 September 2004.

23. Kile, op. cit. note 8, pp. 603, 618.

24. David E. Sanger and William J. Broad, "From Rogue Nuclear Programs, Web of Trails Leads to Pakistan," *New York Times*, 4 January 2004; David E. Sanger and William J. Broad, "As Nuclear Secrets Emerge, More Are Suspected," *New York Times*, 26 December 2004.

25. Kile, op. cit. note 8, p. 603.

26. Ibid., p. 615.

27. Ibid., p. 603.

MAMMALS IN DECLINE (pages 86–87)

1. J. E. M. Baillie, C. Hilton-Taylor, and S. N. Stuart, eds., *2004 IUCN Red List of Threatened Species* (Gland, Switzerland, and Cambridge, U.K.: IUCN–World Conservation Union, 2004).

2. J. Oates, *African Primates: Status Survey and Conservation Action Plan—Revised Edition* (Gland, Switzerland: IUCN, 1996), pp. 78–79; Molly Norton, "Chimpanzees Headed for Extinction," *World Watch*, September/October 2004, p. 10.

3. Baillie, Hilton-Taylor, and Stuart, op. cit. note 1.

4. R. Emslie and M. Brooks, *African Rhino: Status Survey and Conservation Action Plan* (Gland, Switzerland, and Cambridge, U.K.: IUCN, 1999), pp. vii, 5, 11.

5. Ibid.

6. K. Nowell and P. Jackson, *Wild Cats: Status Survey and Conservation Action Plan* (Gland, Switzerland: IUCN, 1996), pp. xii, 175–77; Colby J. Loucks et al., "The Giant Pandas of the Qinling Mountains, China: A Case Study in Designing Conservation Landscapes for Elevational Migrants," *Conservation Biology*, April 2003, pp. 558–65.

7. J. Seidensticker and S. Lumpkin, *Cats: Smithsonian Answer Book* (Washington, DC: Smithsonian Books, 2004), pp. 225–26; Baillie, Hilton-Taylor, and Stuart, op. cit. note 1.

8. R. K. Laidlaw, "Effects of Habitat Disturbance and Protected Areas on Mammals of Peninsular Malaysia," *Conservation Biology*, December 2000, pp. 1639–48.

9. N. J. van Strien, "Conservation Programs for Sumatran and Javan Rhino in Indonesia and Malaysia," in H. Schwammer et al., *Scientific Progress Reports: A Research Update on Elephants and Rhinos: Proceedings of the International Elephant and Rhino Research Symposium, Vienna, June 7-11, 2001* (Münster, Germany: Schüling, 2002), pp. 231–32; L. Naughton-Treves et al., "Wildlife Survival Beyond Park Boundaries: The Impact of Slash-and-Burn Agriculture and Hunting on Mammals in Tambopata, Peru," *Conservation Biology*, August 2003, pp. 1106–17; G. C. Daily et al., "Countryside Biogeography of Neotropical Mammals: Conservation Opportunities in Agricultural Landscapes in Costa Rica," *Conservation Biology*, December 2003, pp. 1814–26.

10. C. Servheen, S. Herrero, and B. Peyton, *Bears: Status Survey and Conservation Action Plan* (Gland, Switzerland, and Cambridge, U.K.: IUCN, 1999), pp. 1–6.

11. Ibid.; Nowell and Jackson, op. cit. note 6, pp. 174, 243.

12. "Food for Thought: The Utilization of Wild Meat in Eastern and Southern Africa," at www.traffic.org/bushmeat/conclusions.html, viewed 28 December 2004; Jennifer Bogo, "Traffic Jam," *Audubon*, March 2003, pp. 102–07.

13. H. T. Dublin et al., *Four Years After the CITES Ban: Illegal Killing of Elephants, Ivory Trade and Stockpiles* (Gland, Switzerland: IUCN/SSC, 1995), pp. 84–87; S. Blake and S. Hedges, "Sinking the Flagship: The Case of Forest Elephants in Asia and Africa," *Conservation Biology*, October 2004, pp. 1191–1202.

14. S. J. Wright et al., "Poachers Alter Mammal Abundance, Seed Dispersal, and Seed Predation in a Neotropical Forest," *Conservation Biology*, February 2000, pp. 227–39; Daily et al., op. cit. note 9.

15. Wright et al., op. cit. note 14.

16. W. J. McShea and J. H. Rappole, "Managing the Abundance and Diversity of Breeding Bird Populations through Manipulation of Deer Populations," *Conservation Biology*, August 2000, pp. 1161–70.

17. Carol Morello, "Whale Deaths Worry Experts," *Washington Post*, 1 December 2004; R. R. Reeves et al., *Guide to Marine Mammals of the World* (New York: Alfred A. Knopf, 2002), pp. 190–93.

18. R. T. T. Forman and Lauren E. Alexander, "Roads and Their Major Ecological Effects," *Annual Review of Ecological Systems*, vol. 29 (1998), pp. 207–31; Laidlaw, op. cit. note 8.

19. Servheen, Herrero, and Peyton, op. cit. note 10, p. 262; Juliet Eilperin, "Study Says Polar Bears Could Face Extinction," *Washington Post*, 9 November 2004.

20. S. D. Williams, "Status and Action Plan for Grevy's Zebra," in P. D. Moehlman, ed., *Equids: Zebras, Asses, and Horses: Status Survey and Conservation Action Plan* (Gland, Switzerland, and Cambridge, U.K.: IUCN, 2002), pp. 11–25.

21. R. Woodroffe and J. R. Ginsberg, "Past and Future Causes of Wild Dogs' Population Decline," *African Wild Dog Status Survey and Action Plan* (Gland, Switzerland: IUCN, 1997), pp. 58–75.

22. J. Short et al., "Surplus Killing by Introduced Predators in Australia: Evidence for Ineffective Anti-predator Adaptations in Native Prey Species?" *Biological Conservation*, vol. 103, no. 3 (2002), pp. 283–301; P. Menkhorst, *A Field Guide to the Mammals of Australia* (New York: Oxford University Press, 2001), pp. 206–07; A. J. Mitchell-Jones et al., *The Atlas of European Mammals* (London: Academic Press, 1999), pp. 178–81, 338–39.

23. A. G. Chiarello, "Density and Population Size of Mammals in Remnants of Brazilian Atlantic Forest," *Conservation Biology*, December 2000, pp. 1649–57.

24. Nowell and Jackson, op. cit. note 6, p. xii.

25. M. G. M. van Roosmalen et al., "A Taxonomic Review of the Titi Monkeys, Genus *Callicebus* Thomas, 1903, with the Description of Two New Species, *Callicebus bernhardi* and *Callicebus stephennashi*, from Brazilian Amazonia," *Neotropical Primates* 10 (Supplement), June 2002, pp. 15–18, 24–30.

26. W. D. Newmark, "Insularization of Tanzanian Parks and the Local Extinction of Large Mammals," *Conservation Biology*, December 1996, pp. 1549–56; D. B. Gurd et al., "Conservation of Mammals in Eastern North American Wildlife Reserves: How Small is Too Small?" *Conservation Biology*, October 2001, pp. 1355–63; G. G. Bruinderink et al.,

"Designing a Coherent Ecological Network for Large Mammals in Northwestern Europe," *Conservation Biology*, April 2003, pp. 549–57; E. Wikramanayake et al., "Designing a Conservation Landscape for Tigers in Human-Dominated Environments," *Conservation Biology*, June 2004, pp. 839–44.

27. Daily et al., op. cit. note 9; Chiarello, op. cit. note 23.

28. Daily et al., op. cit. note 9.

29. Nowell and Jackson, op. cit. note 6, pp. 178, 191.

30. Daily et al., op. cit. note 9; J. A. Hilty and A. M. Merenlender, "Use of Riparian Corridors and Vineyards by Mammalian Predators in Northern California," *Conservation Biology*, February 2004, pp. 126–35.

GLOBAL ICE MELTING ACCELERATING (pages 88–89)

1. Table 1 based on the following: Arctic Climate Impact Assessment (ACIA), *Impacts of a Warming Arctic* (Cambridge: Cambridge University Press, 2004), p. 25; Konrad Steffen and Russell Huff, "A Record Maximum Melt Extent on the Greenland Ice Sheet in 2002," at cires.colorado.edu/steffen/melt, viewed 21 December 2004; Robert W. Corell, Chair, ACIA, Statement before Committee on Commerce, Science, and Transportation, U.S. Senate, Washington, DC, 3 March 2004; David Shukman, "Greenland Ice-Melt 'Speeding Up'," *BBC News Online*, 28 July 2004; Ian Joughin, Waleed Abdalati, and Mark Fahnestock, "Large Fluctuations in Speed on Greenland's Jakobshavn Isbræ Glacier," *Nature*, 2 December 2004, pp. 608–10; Anthony A. Arendt et al., "Rapid Wastage of Alaska Glaciers and Their Contribution to Rising Sea Level," *Science*, 19 July 2002, pp. 382–86; Dan Fagre, "Global Environmental Effects at Glacier National Park," in U.S. National Park Service, *Natural Resource Year in Review 2002* (Washington, DC: U.S. Department of the Interior, 2003), p. 24; Monica Vargas, "Global Warming Melts Peruvian Peaks," *Reuters*, 26 July 2004; Eric Rignot, Andrés Rivera, and Gino Casassa, "Contribution of the Patagonia Icefields of South America to Sea Level Rise," *Science*, 17 October 2003, pp. 434–37; R. Thomas et al., "Accelerated Sea-Level Rise from West Antarctica," *Science*, 8 October 2004, pp. 255–58; Howard W. French, "A Melting Glacier in Tibet Serves as an Example and a Warning," *New York Times*, 9 November 2004; Richard Black, "Climate Change 'Ruining' Everest," *BBC News Online*,

17 November 2004; Mridula Chettri, "Glaciers Beating Retreat," *Down to Earth*, 30 April 1999; Lonnie G. Thompson et al., "Kilimanjaro Ice Core Records: Evidence of Holocene Climate Change in Tropical Africa," *Science*, 18 October 2002, pp. 589–93; European Environment Agency, *Impacts of Europe's Changing Climate* (Copenhagen: 2004), p. 33; Ceri Radford, "Melting Swiss Glaciers Threaten Alps," *Reuters*, 16 November 2004.

2. World Meteorological Organization (WMO), "Global Temperature in 2004 Fourth Warmest," press release (Geneva: 15 December 2004).

3. B. Blair Fitzharris et al., "The Cryosphere: Changes and Their Impacts," in Robert T. Watson et al., eds., *Climate Change 1995. Impacts, Adaptations and Mitigation of Climate Change: Scientific-Technical Analysis* (New York: Cambridge University Press, 1996).

4. ACIA, op. cit. note 1; National Aeronautics and Space Administration, "Seasons of Change: Evidence of Arctic Warming Grows," feature (Washington, DC: 23 October 2003).

5. National Snow and Ice Data Center (NSIDC), "Arctic Sea Ice Decline Continues," *NSIDC News*, 4 October 2004.

6. ACIA, op. cit. note 1.

7. Andrew C. Revkin, "Ice Trends on Greenland," *New York Times*, 14 May 2004.

8. Joughin, Abdalati, and Fahnestock, op. cit. note 1; Waleed Abdalati and Konrad Steffen, "Greenland Ice Sheet Melt Extent: 1979–1999," *Journal of Geophysical Research*, 27 December 2001, pp. 33,983–89.

9. GLACIER, Rice University, "Introduction: How Big is the Ice?" at www.glacier.rice.edu/invitation/1_ice.html, viewed 21 December 2004.

10. "Ice Collapse Speeds Up Glaciers," *BBC News Online*, 22 September 2004.

11. Figure of 13,500 from "Antarctic Glaciers Melting Faster—Study," *Reuters*, 22 September 2004; collapses from NSIDC, "Ice Shelves in the News," at nsidc.org/news/iceshelves; Andrew Shepherd et al., "Larsen Ice Shelf Has Progressively Thinned," *Science*, 31 October 2003, pp. 856–59.

12. Figure of 70 percent from GLACIER, op. cit. note 9.

13. Mike Collett-White, "Ice Shelf Collapse Reignites Global Warming Fears," *Reuters*, 21 March 2002; Eric Rignot et al., "Accelerated Ice Discharge from the Antarctic Peninsula Following the Collapse of Larsen B Ice Shelf," *Geophysical Research Letters* 22 September 2004; Ted A. Scambos et al., "Glacier Acceleration and Thinning After Ice Shelf Collapse

in the Larsen B Embayment, Antarctica," *Geophysical Research Letters*, 22 September 2004.

14. Wilfried Haeberli et al., eds., *Glacier Mass Balance Bulletin, Bulletin No. 7 (2000-2001)* (Zurich: International Commission on Snow and Ice, U.N. Environment Programme (UNEP), UNESCO, and WMO, 2003), p. 82; Alister Doyle, "Glaciers Shrink, But Some Resist Global Warming," *Reuters*, 23 August 2004.

15. Mark Dyurgerov, "Glacier Mass Balance and Regime: Data of Measurements and Analysis," INSTAAR Occasional Paper 55 (Boulder, CO: Institute of Arctic and Alpine Research, 2002), p. 7; equivalencies from Robert S. Boyd, "Scientists Find Dramatic Melting of Glaciers," *Kansas City Star*, 15 August 2002.

16. Fitzharris et al., op. cit. note 3.

17. Laury Miller and Bruce C. Douglas, "Mass and Volume Contributions to Twentieth-Century Global Sea Level Rise," *Nature*, 25 March 2003, pp. 406–09; Arendt et al., op. cit. note 1.

18. ACIA, op. cit. note 1.

19. Michelle C. Mack et al., "Ecosystem Carbon Storage in Arctic Tundra Reduced by Long-Term Nutrient Fertilization," *Nature*, 23 September 2004, pp. 440–43.

20. Ruth Curry, Bob Dickson, and Igor Yashayacv, "A Change in the Freshwater Balance of the Atlantic Ocean Over the Past Four Decades," *Nature*, 18 December 2003, pp. 826–29; Carsten Rühlemann et al., "Warming of the Tropical Atlantic Ocean and Slowdown of Thermohaline Circulation During the Last Deglaciation," *Nature*, 2 December 1999.

21. Andrew Enever, "Bolivian Glaciers Shrinking Fast," *BBC News Online*, 10 December 2002.

22. Navin Singh Khadka, "Himalaya Glaciers Melt Unnoticed," *BBC News Online*, 10 November 2004.

23. Margot Roosevelt, "Vanishing Alaska," *Time*, 4 October 2004; cost of moving from Yereth Rosen, "Warming Climate Disrupts Alaska Natives' Lives," *Reuters*, 19 April 2004.

24. Yana Voitova, "Russian Glacier-Slip Triggers Floods, 95 Missing," *Reuters*, 25 September 2002.

25. UNEP, "Global Warming Triggers Glacial Lakes Flood Threat," press release (Nairobi: 16 April 2002).

26. ACIA, op. cit. note 1.

27. Ibid.; Alister Doyle, "Woes of Warming Arctic to Echo Worldwide Via Birds," *Reuters*, 12 November 2004.

28. "Antarctica Rapidly Getting Greener, Researchers Say," (London) *The Guardian*, 10 September 2003;

Colin Woodard, "For Some Penguins, Less Ice Means Stark Choices," *Christian Science Monitor*, 10 December 1998.

WETLANDS DRYING UP (pages 90–91)

1. Ramsar Convention Secretariat, *The Ramsar Convention Manual: A Guide to the Convention on Wetlands* (Gland, Switzerland: 2004), sections 1.2–1.3.
2. Ibid.
3. C. M. Finlayson and A. G. Spiers, *Global Review of Wetland Resources and Priorities for Wetland Inventory* (Wageningen, The Netherlands: Wetlands International and the Environmental Research Institute of the Supervising Scientist, 1999), section 3.4; World Resources Institute et al., *World Resources 2000–2001* (Washington, DC: 2000), p. 104.
4. Ramsar Convention Secretariat, op. cit. note 1, sections 1.3–1.4.
5. U.N. Environment Programme (UNEP), *Environment in Iraq: UNEP Progress Report* (Geneva: 2003), p. 22; P. Dugan, ed., *Wetlands in Danger: A World Conservation Atlas* (New York: Oxford University Press, 1993), pp. 122–24.
6. H. Partow, *The Mesopotamian Marshlands: Demise of an Ecosystem* (Nairobi: UNEP, 2001), pp. 35–39.
7. Ibid.
8. UNEP, op. cit. note 5, p. 24.
9. Ibid., pp. 5–25.
10. Partow, op. cit. note 6; UNEP, "Will the Aral Sea Disappear Forever?" at www.unep.org/vitalwater/25/htm, viewed 10 January 2005; Dugan, op. cit. note 5, p. 149.
11. L. V. Balian et al., "Changes in the Waterbird Community of the Lake Sevan–Lake Gilli Area, Republic of Armenia: A Case for Restoration," *Biological Conservation*, August 2002, pp. 157–63.
12. UNEP, "Wetland Shared Between Afghanistan, Iran Almost Completely Dry, According to Report Presented to Environment Leaders in Nairobi," press release (Nairobi: 6 February 2003).
13. Partrow, op. cit. note 6.
14. P. S. Goodman, "Manipulating the Mekong," *Washington Post*, 30 December 2004.
15. M. Alvarez-Cobelas, S. Cirujano, and S. Sanchez-Carillo, "Hydrological and Botanical Man-made Changes in the Spanish Wetland of Las Tablas de Daimiel," *Biological Conservation*, January 2001, pp. 89–98.
16. Ibid.
17. Ibid.
18. Chris Bright, *Life Out of Bounds* (New York: W.W. Norton & Company, 1998), pp. 89–91, 182–83.
19. T. Chea, "Marsh Grass Threatens Plants in S.F. Bay," *Associated Press*, 10 November 2004; George W. Cox, *Alien Species in North America and Hawaii* (Washington, DC: Island Press, 1999), pp. 57, 270–71.
20. Cox, op. cit. note 19, pp. 199, 229–30.
21. Alvarez-Cobelas, Cirujano, and Sanchez-Carillo, op. cit. note 15.
22. Ramsar Convention Secretariat, op. cit. note 1, sections 1.4–1.7.5; Wetlands International, "Myanmar Joins the Ramsar Convention," press release (Wageningen, The Netherlands: 16 December 2004).
23. Finlayson and Spiers, op. cit. note 3, section 3.1.
24. Ibid.
25. Ibid., section 3.5.
26. T. E. Dahl, *Status and Trends of Wetlands in the Conterminous United States, 1986 to 1997* (Washington, DC: U.S. Department of the Interior, Fish and Wildlife Service, 2000), pp. 10–11, 45.
27. Ibid.
28. T. E. Dahl, *Wetlands: Losses in the United States, 1780's to 1980's* (Washington, DC: U.S. Department of the Interior, Fish and Wildlife Service, 1990), p. 1.

FOREST LOSS CONTINUES (pages 92–93)

1. Dirk Bryant, Daniel Nielsen, and Laura Tangley, *Last Frontier Forests: Ecosystems and Economies on the Edge* (Washington, DC: World Resources Institute, 1997), p. 6.
2. U.N. Food and Agriculture Organization (FAO), *Global Forest Resources Assessment 2000* (Rome: 2000), Executive Summary.
3. Ibid.
4. Ibid.
5. Alex de Sherbinin, *A Guide to Land Use and Land Cover Change* (Palisades, NY: CIESIN, Columbia University, 2002), p. 1.
6. "Deforestation Doubles in Indonesia," at www.peopleandplanet.net, 7 March 2002.
7. "Amazon Forest Still Shrinking Fast," at www.peopleandplanet.net, 8 April 2004.
8. Ibid.
9. Worldwatch calculation based on Intergovernmental Panel on Climate Change, *Land Use, Land-Use Change, and Forestry*, Summary for Policymakers (Nairobi: U.N. Environment Programme, 2000), p. 4.
10. "Deforestation Doubles in Indonesia," op. cit. note 6.

11. "Amazon Forest Still Shrinking Fast," op. cit. note 7.

12. Amy Bracken, "Deadly Floods in Haiti Blamed on Deforestation, Poverty," *Environmental News Network*, 23 September 2004.

13. "Colombia, of All Places, Facing Water Shortages," at www.ecoamericas.com, December 2004.

14. Bracken, op. cit. note 12.

15. Helmut J. Geist and Eric F. Lambin, "What Drives Tropical Deforestation?" LUCC Report Series No. 4 (Louvain-la-Neuve, Belgium: Land Use and Land Cover Change Project, 2001), p. 1.

16. Ibid.

17. Ibid.

18. Ibid.

19. Ibid., pp. 24, 27.

20. Ibid.

21. Ibid., p. 95.

22. "Deforestation Doubles in Indonesia," op. cit. note 6.

23. Ibid.

24. FAO, *State of the World's Forests 2003*, Institutional Framework (Rome: 2003).

25. Ibid.

26. Forest Stewardship Council, at www.certified-forests .org/data/increase.htm, viewed 1 February 2005.

27. FAO, op. cit. note 2, Chapter 6.

28. "The World Bank/WWF Forest Alliance," at lnweb18 .worldbank.org/ESSD/envext.nsf/80ByDocName/ WBWWFForestAlliance, viewed 9 February 2005.

AIR POLLUTION STILL A PROBLEM (pages 94–95)

1. World Health Organization and European Environment Agency, *Air and Health—Local Authorities, Health and Environment* (Copenhagen: 1997), p. 3.

2. World Bank, *World Development Indicators 2003* (Washington, DC: 2003).

3. Worldwatch calculation based on data in ibid., pp. 168–69.

4. Ibid.

5. Ibid.

6. National Institute of Environmental Health Sciences, "Link Strengthened Between Lung Cancer, Heart Deaths and Tiny Particles of Soot, Dust," press release (Triangle Park, NC: 5 March 2002).

7. Ibid.; Ben Harder, "No Deep Breathing: Air Pollution Impedes Lung Development," *Science News*, 11 September 2004, p. 163.

8. Rob McConnell et al., "Asthma in Exercising Children Exposed to Ozone," *The Lancet*, 2 February 2002, pp. 386–91.

9. Ibid.

10. Luis Cifuentes et al., "Hidden Health Benefits of Greenhouse Gas Mitigation," *Science*, 17 August 2001, pp. 1257–59.

11. U.N. Environment Programme (UNEP), "Over 50% of Gasoline in Sub-Saharan Africa is Now Lead Free," press release (Nairobi: 7 May 2004).

12. World Bank, *Fuel for Thought: An Environmental Strategy for the Energy Sector* (Washington, DC: June 2000), p. 98.

13. European Environment Agency, *Air and Health—Air Pollution and Global Effects* (Copenhagen: 1997), p. 9.

14. UNEP and the Center for Clouds, Chemistry, and Climate, *The Asian Brown Cloud: Climate and Other Environmental Impacts* (Klong Luang, Thailand: Regional Resource Centre for Asia and the Pacific, UNEP, August 2002), p. 44.

15. Ibid.

16. Mike Holland et al., *Economic Assessment of Crop Yield Losses from Ozone Exposure* (UK: Centre for Ecology and Hydrology, April 2002), pp. 1–2.

17. Fred Pearce, "Smog Crop Damage Costs Billions," *New Scientist*, 11 June 2002.

18. Valerie M. Thomas et al., "Effects of Reducing Lead in Gasoline," *Environmental Science and Technology*, 15 November 1999, pp. 3942–48.

19. Cifuentes et al., op. cit. note 10.

20. European Automobile Manufacturers Association, "World's Automakers Promote Global Phaseout of Leaded Gasoline in World-Wide Fuel Charter, Recommend Completing by 2005," press release (Brussels: 30 August 2002).

21. UNEP, op. cit. note 11.

22. Ibid.

23. Committee on Environmental Health, American Academy of Pediatrics, "Ambient Air Pollution: Health Hazards to Children," *Pediatrics*, December 2004, pp. 1699–1707.

SOCIALLY RESPONSIBLE INVESTING SPREADS (pages 98–99)

1. Worldwatch calculation based on Social Investment Forum (SIF), *2003 Report on Socially Responsible Investing Trends in the United States* (Washington, DC: 2003), on Eurosif, *Socially Responsible Investment Among European Institutional Investors, 2003 Report* (Paris: 2003), on Social Investment Organization (SIO), *Canadian Social Investment Review 2002* (Toronto: 2003), on Ethical Investment

Association (EIA), *Socially Responsible Investment in Australia–2003* (Sydney: 2003), and on International Finance Corporation (IFC), *Towards Sustainable and Responsible Investment in Emerging Markets* (Washington, DC: 2003); Canadian estimate is for 2002; European estimate includes the United Kingdom, the Netherlands, France, Spain, Italy, Germany, Austria, and Switzerland.

2. SIF, op. cit. note 1, p. ii.
3. Eurosif, op. cit. note 1, p. 10.
4. SIO, op. cit. note 1, p. 5; EIA, op. cit. note 1, p. 1; SIF, op. cit. note 1, p. 33; IFC, op. cit. note 1, p. 18.
5. SIF, op. cit. note 1, p. 3.
6. KLD Research and Analytics, Inc., "The Domini 400 Social Index," at www.kld.com/benchmarks/dsi.html, viewed 1 February 2005; KLD Research and Analytics, Inc., "Composition of the Index," www.kld.com/benchmarks/dsicomposition.html, viewed 1 February 2005.
7. Marshall Glickman and Marjorie Kelly, "Working Capital: Can Socially Responsible Investing Make a Great Green Leap Forward?" *E Magazine*, March/April 2004, pp. 26–38.
8. Eurosif, op. cit. note 1, p. 6.
9. Glickman and Kelly, op. cit. note 7.
10. SIF, op. cit. note 1, p. 2.
11. Ibid., p. ii; SIF, *2001 Report on Socially Responsible Investing Trends in the United States* (Washington, DC: 2001), p. 5. In Figure 1, no breakdown is available for 1984, and estimates do not include community investing.
12. SIF, op. cit. note 1, p. 1.
13. Ibid., p. 2.
14. Ibid.
15. Ibid., pp. 2, 18.
16. Northwest Corporate Accountability Project, "What Is a Shareholder Resolution?" at www.scn.org/earth/wum/2Whatsr.htm, viewed 1 February 2005.
17. SIF, op. cit. note 1, p. 18.
18. Ibid., p. 20.
19. Ibid., p. 23.
20. Ibid., pp. 24–25.
21. CDFI Data Project, *CDFIs: Providing Capital, Building Communities, Creating Impact* (Arlington, VA: 2002), p. 4.
22. Eurosif, op. cit. note 1, p. 10.
23. Ibid.
24. Ibid., p. 11.
25. Ibid. p. 18.
26. Ibid.

27. Ibid., p. 11; SIF, op. cit. note 1, p. 31.
28. SIF, op. cit. note 1, p. 32.
29. Ibid.
30. Ibid.
31. IFC, op. cit. note 1, p. 18.
32. Ibid., pp. 25–28.
33. Martin Wright, "Spiritual Capital Across the Divide," *Green Futures*, 25 March 2003.
34. Paul Vallely, "How to Do Business for Good," *Church Times*, 22 November 2002; International Interfaith Investment Group (3iG), "The Definition of a Good Investment," brochure (Manchester, U.K.: no date).
35. 3iG, op. cit. note 34.

INTEREST IN RESPONSIBLE TRAVEL GROWS
(pages 100–101)

1. Definition from Merriam Webster's Dictionary; arrivals from World Tourism Organization, *World Tourism Barometer*, January 2005, p. 2. International tourism arrivals measure the number of people passing through immigration; an individual could therefore be counted more than once in a year.
2. World Tourism Organization, op. cit. note 1, p. 4.
3. World Tourism Organization, *Tourism Highlights: Edition 2004* (Madrid: 2004).
4. World Tourism Organization, op. cit. note 1, p. 3.
5. World Travel & Tourism Council (WTTC), *Travel & Tourism Forging Ahead: Executive Summary* (London: 2004), p. 4.
6. Ibid.
7. Office of the High Representative for the Least Developed Countries, Landlocked Developing Countries and Small Island States, "List of Least Developed Countries," fact sheet, updated December 2004; WTTC, op. cit. note 5, pp. 4, 23.
8. U.N. Environment Programme, "Economic Impacts of Tourism," at www.uneptie.org/pc/tourism/sust-tourism/economic.htm.
9. Ibid.
10. Rebecca Bowe et al., "Flying the Dirty Skies," *E Magazine*, September/October 2004; "Indian Budget Airline to Expand," *BBC News*, 10 May 2004.
11. Companies offering this option include Climate Care, at www.co2.org, and Future Forests, at www.futureforests.com.
12. Estimates based on average calculations from three carbon offset programs: Climate Care, op. cit. note 11; Future Forests, op. cit. note 11; and myclimate, at www.myclimate.org.

13. International Hotels Environment Initiative, "Consumer Attitudes Towards the Role of Hotels in International Environmental Sustainability," press release (London: 23 July 2002).

14. Definitions from Merriam-Webster Dictionary, *National Geographic Traveler*, International Ecotourism Society, World Tourism Organization, Pro-Poor Tourism, Responsibletravel.com, and David B. Weaver, ed., *Encyclopedia of Ecotourism* (New York: CABI, 2001), p. 659.

15. Martha Honey, The International Ecotourism Society, discussion with author, 2 February 2005.

16. Martha Honey, "An Overview of Certification," presentation to the World Parks Congress, Durban, South Africa, 12 September 2003.

17. Ronald Sanabria, Rainforest Alliance, e-mail to author, 12 February 2005.

18. Africa Foundation, "Our Achievements," fact sheet, at www.africafoundation.org, viewed 10 February 2005.

19. Ibid.

20. Airline Ambassadors International, "About Us," fact sheet, at www.airlineamb.org, viewed 10 February 2005.

GLOBAL JOBS SITUATION STILL POOR
(pages 102–103)

1. International Labour Organization (ILO), *Global Unemployment Trends 2004* (Geneva: 2004).

2. ILO, "Half the World's Workers Living Below US$2 a Day Poverty Line," press release (Geneva: 7 December 2004).

3. ILO, *World Employment Report 2004/2005* (Geneva: 2004), p. 24.

4. The ILO defines the unemployed as members of the economically active population who are currently without work, who are seeking or have sought work recently, and who are currently available for work. ILO, op. cit. note 1.

5. Unemployment rates derived from ILO, op. cit. note 1, from ILO, *Global Employment Trends for Youth 2004* (Geneva: 2004), p. 8, and from ILO, Bureau of Statistics, "Labour Market and Economic Indicators, Selected Years" (table), at www.ilo.org/public/english/bureau/inf/images/wer2004/table.gif.

6. ILO, op. cit. note 1.

7. ILO, op. cit. note 3, p. 25.

8. Ibid., p. 67.

9. Ibid., p. 25.

10. ILO, "ILO Calls for Integrated Employment Strategy for Tsunami Reconstruction, Says Estimated 1 Million People Lost Livelihood," press release (Geneva: 19 January 2005).

11. Ibid.

12. ILO, op. cit. note 3, p. 69.

13. Ibid., p. 6; 6 percent from ILO, op. cit. note 1.

14. ILO, *Key Indicators of the Labour Market, 3rd Edition* (Geneva: 2004).

15. Ibid.

16. ILO, "New ILO Book Explores 'Decent Working Time Deficit' in the Industrialized Countries," press release (Geneva: 22 October 2004).

17. U.S. estimates cited in ILO, op. cit. note 3, p. 70.

18. World Bank, *World Development Indicators 2004* (Washington, DC: 2004), p. 53.

19. Alejandro Portes, Manuel Castells, and Lauren A. Benton, *The Informal Economy: Studies in Advanced and Less Developed Countries* (Baltimore: Johns Hopkins University Press, 1989).

20. National Sample Survey Organisation, Government of India, *Employment and Unemployment Situation in India, 1999–2000* (Calcutta: 2001); Ministry of Labour, Government of India, *Report of the 2nd National Commission on Labour* (New Delhi: 2002).

21. World Bank, op. cit. note 18, p. 45.

22. ILO, op. cit. note 3, p. 25.

23. Ibid.

24. ILO, *Global Employment Trends for Women 2004* (Geneva: 2004).

25. World Bank, op. cit. note 18, p. 49.

26. ILO, *Global Employment Trends for Youth 2004*, op. cit. note 5; 6.2 percent from ILO, "Global Unemployment Remains at Record Levels in 2003 But Annual ILO Jobs Report Sees Signs of Recovery," press release (Geneva: 22 January 2004).

27. Richard P. Cincotta, Robert Engelman, and Daniele Anastasion, *The Security Demographic: Population and Civil Conflict After the Cold War* (Washington, DC: Population Action International, 2003); Jack A. Goldstone, "Population and Security: How Demographic Change Can Lead to Violent Conflict," *Journal of International Affairs*, fall 2002, pp. 3–22.

28. Estimate of 40 percent from ILO, op. cit. note 3, p. 14; Leif Ohlsson, *Livelihood Conflicts—Linking Poverty and Environment as Causes of Conflict* (Stockholm: Swedish International Development Cooperation, Environmental Policy Unit, 2000).

29. Chris Dolan, "Collapsing Masculinities and Weak States—A Case Study of Northern Uganda," in Frances Cleaver, ed., in *Masculinities Matter!* (London: Zed Books, April 2003), pp. 57–83.

30. ILO, op. cit. note 3, p. 55–56.

31. ILO, op. cit. note 2.

32. ILO, op. cit. note 3, pp. 31–32.

GLOBAL PUBLIC POLICY COOPERATION GROWS (pages 106–107)

1. The term was first coined by Wolfgang Reinicke; see, for example, Wolfgang Reinicke, "Global Public Policy," *Foreign Affairs*, November/December 1997, pp. 127–38.

2. Wolfgang Reinicke et al., *Critical Choices. The United Nations, Networks, and the Future of Global Governance* (Ottawa, ON, Canada: International Development Research Centre, 2000).

3. Jan Martin Witte, Wolfgang H. Reinicke, and Thorsten Benner, "Beyond Multilateralism: Global Public Policy Networks," *International Politics and Society*, no. 2/2000.

4. Jeffry A. Frieden and David A. Lake, *International Political Economy: Perspectives on Global Power and Wealth* (New York: St. Martin's Press, 1999), p. 284.

5. Hilary French, Gary Gardner, and Erik Assadourian, "Laying the Foundations for Peace," in Worldwatch Institute, *State of the World 2005* (New York: W.W. Norton & Company, 2005), p. 173.

6. Consultative Group on International Agricultural Research, at www.cgiar.org.

7. French, Gardner, and Assadourian, op. cit. note 5, p. 176.

8. African Stockpiles Program, at www.africastock piles.org.

9. Charlotte Streck, "Global Public Policy Networks as Coalitions for Change," in Daniel C. Esty and Maria H. Ivanova, eds., *Global Environmental Governance: Options & Opportunities* (New Haven: Yale School of Forestry and Environmental Studies, 2002), pp. 121–40.

10. International Forum on Forests, at www.un.org/esa/forests/ipf_iff.html.

11. Global Water Partnership, at www.gwpforum.org.

12. J. F. Rischard, *High Noon: 20 Global Problems, 20 Years to Solve Them* (New York: Basic Books, 2002), p. 184.

13. Millennium Ecosystem Assessment (MA), at www.millenniumassessment.org.

14. Streck, op. cit. note 9.

15. MA, op. cit. note 13.

GREATER EFFORT NEEDED TO ACHIEVE THE MDGS (pages 108–109)

1. This vital sign draws heavily on research and writing originally done by Erik Assadourian for the Institute's *State of the World 2005* report. "World Leaders Adopt 'United Nations Millennium Declaration' at Conclusion of Extraordinary Three-Day Summit," press release (New York: United Nations, 8 September 2000); Box 1 from United Nations, Millennium Development Goals, available at www.un.org/millenniumgoals.

2. Annan quote from U.N. General Assembly, "Report of the Secretary-General on the Modalities, Format and Organization of the High-level Plenary Meeting of the 60th Session of the General Assembly," New York, p. 5.

3. U.N. Development Programme (UNDP), *Human Development Report 2004* (New York: Oxford University Press, 2004), p. 133.

4. World Bank, *Global Monitoring Report 2004* (Washington, DC: 2004), pp. 41–45.

5. UN Millennium Project, *Investing in Development: A Practical Plan to Achieve the Millennium Development Goals—Overview* (New York: UNDP, 2005); World Economic Forum, *Global Governance Initiative 2005*, *Executive Summary* (Cologny/Geneva: 2005).

6. UN Millennium Project, op. cit. note 5, pp. 25–27.

7. Alex Bellos, "Brazil's New Leader Shelves Warplanes to Feed Hungry," *The Guardian* (London), 4 January 2003.

8. UNDP, *Human Development Report 2003* (New York: Oxford University Press, 2003), p. 98; UNDP, op. cit. note 3, pp. 139–42.

9. World Health Organization estimates cited in UN Millennium Project, *Investing in Development: A Practical Plan to Achieve the Millennium Development—Full Report*, available at unmp.forumone.com/eng_full_report/MainReportComplete-lowres.pdf, p. 60.

10. Gross domestic product (GDP) in low-income countries from U.N. Conference on Trade and Development (UNCTAD), *The Least Developed Countries Report 2004, Statistical Annex* (New York: United Nations, 2004), p. 321.

11. Development Assistance Committee (DAC), *ODA Statistics for 2003 and ODA Outlook* (Paris: Organisation for Economic Co-operation and Development,

2004), p. 4; gross national income is a similar measure to GDP but adds income earned abroad by companies of the country referenced while subtracting income earned by foreign companies in the referenced country; commitments to meet 0.7 percent target from UN Millennium Project, op. cit. note 9, p. 59, and from United Nations, *Johannesburg Declaration on Sustainable Development and Plan of Implementation of the World Summit on Sustainable Development* (New York: United Nations, 2003), p. 65.

12. DAC, op. cit. note 11, p. 4.

13. Official development assistance needs from UN Millennium Project, op. cit. note 9, pp. 55–60.

14. Tied aid from UNDP, *Development Effectiveness Report 2003: Partnerships for Results* (Washington, DC: 2003), p. 41; aid for services from IBON Foundation, *The Reality of Aid 2004: Focus on Governance and Human Rights in International Cooperation* (Manila: 2004), pp. 181–94, and from Santosh Mehrotra, *The Rhetoric of International Development Targets and the Reality of Official Development Assistance* (Florence, Italy: UNICEF Innocenti Research Centre, 2001).

15. U.S. aid spending on Iraq reconstruction from Curt Tarnoff and Larry Nowels, "Foreign Aid: An Introductory Overview of U.S. Program and Policies," Congressional Research Service, Library of Congress, updated 15 April 2004, p. 13.

16. Calculated from data presented in Tarnoff and Nowels, op. cit. note 15, pp. 13, 29; poorest countries from UNCTAD, op. cit. note 10, p. 318.

17. U.N. Statistics Division, *Progress Towards the Millennium Development Goals, 1990–2003*, unofficial working paper (New York: 23 March 2004).

The Vital Signs Series

Some topics are included each year in *Vital Signs*; others are covered only in certain years. The following is a list of topics covered in *Vital Signs* thus far, with the year or years they appeared indicated in parentheses. Those marked with a bullet (•) appeared in Part One, which includes time series of data on each topic.

AGRICULTURE AND FOOD

Agricultural Resources
- Fertilizer Use (1992–2001)
- Grain Area (1992–93, 1996–97, 1999–2000)
- Grain Yield (1994–95, 1998)
- Irrigation (1992, 1994, 1996–99, 2002)
  Livestock (2001)
  Organic Agriculture (1996, 2000)
  Pesticide Control or Trade (1996, •2000, 2002)
  Transgenic Crops (1999–2000)
  Urban Agriculture (1997)

Food Trends
- Aquaculture (1994, 1996, 1998, 2002, 2005)
  Biotech Crops (2001–02)
- Cocoa Production (2002)
- Coffee (2001)
- Fish (1992–2000)
- Grain Production (1992–2003, 2005)
- Grain Stocks (1992–99)
- Grain Used for Feed (1993, 1995–96)
- Meat (1992–2000, 2003, 2005)
- Milk (2001)
- Soybeans (1992–2001)
- Sugar and Sweetener Use (2002)

THE ECONOMY

Resource Economics
  Agricultural Subsidies (2003)
- Aluminum (2001)
  Arms and Grain Trade (1992)
  Commodity Prices (2001)
  Fossil Fuel Subsidies (1998)
- Gold (1994, 2000)
  Illegal Drugs (2003)
  Metals Exploration (1998, •2002)
- Metals Production (2002)
- Paper (1993, 1994, 1998–2000)
  Paper Recycling (1994, 1998, 2000)
- Roundwood (1994, 1997, 1999, 2002)
  Seafood Prices (1993)
- Steel (1993, 1996, 2005)
  Steel Recycling (1992, 1995)
  Subsidies for Environmental Harm (1997)
  Wheat/Oil Exchange Rate (1992–93, 2001)

Natural Resource Uses
Biomass Energy (1999)
Dams (1995)
Ecosystem Conversion (1997)
Energy Productivity (1994)
Organic Waste Reuse (1998)
Soil Erosion (1992, 1995)
Tree Plantations (1998)

Pollution
Acid Rain (1998)
Air Pollution (1993, 1999, 2005)
Algal Blooms (1999)
Hazardous Wastes (2002)
Lead in Gasoline (1995)
Nuclear Waste (1992, •1995)
Pesticide Resistance (•1994, 1999)
• Sulfur and Nitrogen Emissions (1994–97)

Other Environmental Topics
Environmental Treaties (•1995, 1996, 2000, 2002)
Nitrogen Fixation (1998)
Pollution Control Markets (1998)
Sea Level Rise (2003)
Semiconductor Impacts (2002)
Transboundary Parks (2002)
• World Heritage Sites (2003)

THE MILITARY
• Armed Forces (1997)
Arms Production (1997)
• Arms Trade (1994)
Landmines (1996, 2002)
• Military Expenditures (1992, 1998, 2003, 2005)
• Nuclear Arsenal (1992–96, 1999, 2001, 2005)
Peacekeeping Expenditures (1993, •1994–2003, •2005)
Resource Wars (2003)
• Wars (1995, 1998–2003, 2005)
Small Arms (1998–99)

SOCIETY AND HUMAN WELL-BEING

Health
• AIDS/HIV Incidence (1994–2003, 2005)
Alternative Medicine (2003)
Asthma (2002)
Breast and Prostate Cancer (1995)
• Child Mortality (1993)
• Cigarettes (1992–2001, 2003, 2005)
Drug Resistance (2001)
Endocrine Disrupters (2000)
Food Safety (2002)
Hunger (1995)
• Immunizations (1994)
• Infant Mortality (1992)
Infectious Diseases (1996)
Life Expectancy (1994, •1999)
Malaria (2001)
Malnutrition (1999)
Mental Health (2002)
Mortality Causes (2003)
Noncommunicable Diseases (1997)
Obesity (2001)
• Polio (1999)
Safe Water Access (1995)
Sanitation (1998)
Soda Consumption (2002)
Traffic Accidents (1994)
Tuberculosis (2000)

Reproduction and Women's Status
Family Planning Access (1992)
Female Education (1998)
Fertility Rates (1993)
Maternal Mortality (1992, 1997, 2003)
• Population Growth (1992–2003, 2005)
Sperm Count (1999)
Violence Against Women (1996, 2002)
Women in Politics (1995, 2000)

Social Inequities
Homelessness (1995)
Income Distribution (1992, 1995, 1997, 2002–03)

Language Extinction (1997, 2001)
Literacy (1993, 2001)
Prison Populations (2000)
Social Security (2001)
Teacher Supply (2002)
Unemployment (1999, 2005)

Other Social Topics

Aging Populations (1997)
Fast-Food Use (1999)
International Criminal Court (2003)
Millennium Development Goals (2005)
Nongovernmental Organizations (1999)
Orphans Due to AIDS Deaths (2003)
Public Policy Networks (2005)
Refugees (•1993–2000, 2001, 2003,
    •2005)
Religious Environmentalism (2001)
Urbanization (•1995–96, •1998, •2000,
    2002)
Voter Turnouts (1996, 2002)
Wind Energy Jobs (2000)

TRANSPORTATION AND COMMUNICATIONS

•Air Travel (1993, 1999, 2005)
•Automobiles (1992–2003, 2005)
•Bicycles (1992–2003, 2005)
 Car-Sharing (2002)
 Computer Production and Use (1995)
 Gas Prices (2001)
 Electric Cars (1997)
•Internet (1998–2000, 2002)
•Internet and Telephones Combined
    (2003)
•Motorbikes (1998)
•Railroads (2002)
•Satellites (1998–99)
•Telephones (1998–2000, 2002)
 Urban Transportation (1999, 2001)

Worldwatch Publications

State of the World 2005
REDEFINING GLOBAL SECURITY

Worldwatch's flagship annual is used by government officials, corporate planners, journalists, development specialists, professors, students, and concerned citizens in over 120 countries. Published in more than 20 different languages, it is one of the most widely used resources for analysis. The authors of *State of the World 2005* propose that the foundations for peace and stability lie in moving away from dependence on oil, managing water conflicts, containing infectious diseases, moving toward disarmament, cultivating food security, and cooperating across borders to achieve a sustainable world.

Vital Signs 2005

From shrinking forests to rising prosperity in China, *Vital Signs 2005* documents the trends that are shaping our future in concise analyses and clear tables and graphs.

This thirteenth volume of the Worldwatch Institute series finds that true global progress cannot be achieved as long as the world's spending priorities are directed toward narrow economic and military goals at the expense of human development and environmental protection.

State of the World 2005 Library

Subscribe to the *State of the World Library* and join thousands of decisionmakers and concerned citizens who stay current on emerging environmental issues. For 2005 the Library includes our flagship annual *State of the World 2005: Redefining Global Security, Vital Signs 2005*, a Worldwatch Paper on freshwater ecosystems, and a new Worldwatch Book on religion and sustainability.

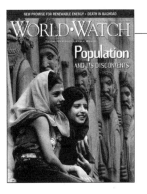

WORLD·WATCH

This award-winning bimonthly magazine is internationally recognized for the clarity and comprehensiveness of its articles on global trends. Keep up to speed on the latest developments in population growth, climate change, species extinction, and the rise of new forms of human behavior and governance. There is no other magazine like it in the world.

Signposts 2004

More comprehensive than ever! Includes 238 datasets of global trends (100 brand new and 138 updated datasets). Each dataset is accompanied by PowerPoint slides of charts and graphs, Excel files, and HTML pages—ready for your use in classroom and boardroom presentations. Also includes full text of *State of the World 2001, 2002, 2003,* and *2004* and *Vital Signs 2001, 2002,* and *2003,* with more than 50 years of environmental, economic, and social indicators, plus an electronic timeline with links to 449 web resources.

Worldwatch Book Series

In *Eat Here,* learn why eating local food is one of the most significant choices you can make for the planet and yourself. Discover why local food products are better for your health, farmers, and the environment. Find out why long-distance food can be dangerous. Get practical advice on finding homegrown pleasures in an anonymous food chain.

Four Easy Ways to Order

❶ Call us at 888-544-2303 or 570-320-2076
❷ Fax us at 570-320-2079
❸ E-mail us at wwpub@worldwatch.org
❹ Visit us on the Web at www.worldwatch.org

Worldwatch Papers

Worldwatch Papers are written by the same award-winning team that produces *State of the World*. Each 50–70 page *Paper* provides cutting-edge analysis on an environmental topic that is making—or is about to make—headlines worldwide. Selected available *Papers* appear by topic below.

To order these and other Worldwatch publications, call us at 888-544-2303 or 570-320-2076, fax us at 570-320-2079, e-mail us at wwpub@worldwatch.org, or visit our Web site at www.worldwatch.org.

About the Worldwatch Institute

RESEARCH PROGRAMS

The Worldwatch Institute's interdisciplinary approach allows its team of researchers to explore emerging global issues from many perspectives, drawing on insights from ecology, economics, public health, sociology, and a range of other disciplines. The Institute's four research teams focus on:

- People
- Nature
- Energy
- Economy

PRESS INQUIRIES

Worldwatch provides reporters from around the world with access to the Institute's extensive research and the researchers behind it. For current information available to the media, visit our online press center at www.worldwatch.org/press.

For press inquiries or to be placed on the Worldwatch media list, contact Darcey Rakestraw by phone at 202-452-1992, ext. 517, by fax at 202-296-7365, or by e-mail at drakestraw@worldwatch.org.

SPEAKERS BUREAU

Worldwatch researchers have extensive experience in bringing audiences up to date on important global trends, including food, water, pollution, climate, forests, oceans, energy, technology, and environmental security.

For more information, or to schedule a speaker, call Gary Gardner at 202-452-1992, ext. 521, or e-mail: ggardner@worldwatch.org.

INTERNATIONAL PUBLISHING PROGRAM

Worldwatch works with overseas publishers to translate, produce, and market its books, papers, and magazine. The Institute has more than 160 publishing contracts in over 20 languages. A complete listing can be found at www.worldwatch.org/foreign/index.html.

For more information, contact Patricia Shyne by phone at 202-452-1992, ext. 520, by fax at 202-296-7365, or by e-mail at pshyne@worldwatch.org.

WORLDWATCH ONLINE

The Worldwatch Web site (www.worldwatch.org) provides immediate access to the Institute's publications. Save time and money by ordering and downloading Worldwatch publications in pdf format from our online bookstore. The site also includes press releases, special briefings on breaking environmental news, contact information, and job announcements.

SUBSCRIBE TO WORLDWATCH NEWS

Worldwatch maintains a free one-way e-mail list to distribute updates from the Institute as well as press releases on new books, papers, and magazine articles.

To subscribe, visit the Worldwatch Web site at www.worldwatch.org.

FRIENDS OF WORLDWATCH

The Worldwatch Institute is a 501 (c)(3) non-profit organization. We rely on gifts from individuals and foundations to underwrite our efforts to provide the information and analysis needed to foster an environmentally sustainable society.

Your gift will be used to help Worldwatch broaden its outreach programs to decisionmakers, build relationships with overseas environmental groups, and disseminate its vital information to as many people as possible through the Institute's Web site and publications.

To join our family of supporters, please call us at 202-452-1992, ext. 530. You can also donate online at www.worldwatch.org/donate.

LEGACY FOR SUSTAINABILITY

You can make a lasting contribution to a better future by remembering Worldwatch in your will. If you are interested in naming the Institute in your will, please contact us.

For further information on giving to Worldwatch, please contact John Holman by phone at 202-452-1992, ext. 518, by fax at 202-296-7365, or by e-mail at jholman@worldwatch.org.

Worldwatch Institute

VISION FOR A SUSTAINABLE WORLD

The WORLDWATCH INSTITUTE is an independent research organization that works for an environmentally sustainable and socially just society, in which the needs of all people are met without threatening the health of the natural environment or the well-being of future generations.

By providing compelling, accessible, and fact-based analysis of critical global issues, Worldwatch informs people around the world about the complex interactions between people, nature, and economies. Worldwatch focuses on the underlying causes of and practical solutions to the world's problems, in order to inspire people to demand new policies, investment patterns, and lifestyle choices.